ENVIRONMENT, DEVELOPMENT AND HUMAN SECURITY

Perspectives from South Asia

Edited by
Adil Najam

University Press of America,® Inc.
Lanham · New York · Oxford

University Press of America,® Inc.
4501 Forbes Boulevard
Suite 200
Lanham, Maryland 20706
UPA Acquisitions Department (301) 459-3366

PO Box 317
Oxford
OX2 9RU, UK

Printed in the United States of America
British Library Cataloging in Publication Information Available

ISBN 0-7618-2571-1 (paperback : alk. ppr.)

∞™ The paper used in this publication meets the minimum requirements of American National Standard for Information Sciences—Permanence of Paper for Printed Library Materials, ANSI Z39.48—1984

Contents

Boxes, Figures & Tables

Boxes

Figures

Tables

Foreword

AMONG THE MANY transformations we are witnessing at the dawn of the 21st Century, a crucial issue is one related to the concept of security. The fundamental question that arises is what is it we want to secure? In the earlier era of strong state nationalism, the answer was obvious. It was the territory of the nation state and as a corollary all that furthered a nation's interests in this task. This could be achieved essentially at the expense of another country or a group of states. This idea held sway till the end of the Second World War and continued for much of the Cold War. It was common then, in the name of this nationalism, to expect citizens of a state to unquestionably lay down their lives in defense of this cause. This no longer holds. Countries today find it difficult to justify the loss of a single citizen's life in conflicts, whether near or far from home. Because increasingly in people's perceptions these have slipped in priority compared to other issues. What then is security?

In truth, security is far too interdependent an issue to be defined in purely military terms. In a rapidly integrating and globalizing world, other factors and concerns have come to acquire greater salience. It can no longer be my security against yours, but my security *and* your security, in a single planet where all are equal stakeholders. Full realization of this and a clearer definition will perhaps take some more time yet, but the concept and relevance of *human security* vis a vis *state security* is growing and entering the international debate.

There is now a growing salience in people's minds on three fundamental human aspirations; *freedom from fear*, *freedom from want* and the hope of leading a *life of dignity*. These are challenged today not so much by war or conflict, but other perhaps even more serious and immediate concerns. Three of the more important issues particularly in Asia are; the impact of globalization particularly on developing nations, the challenges to governance in plural societies and the adverse impact of a degrading environment.

It was with this aim and with support from the Ford Foundation, that the Regional Centre for Strategic Studies initiated a project on Non Traditional Security Issues in South Asia. This was an Asia-wide program examining these emerging conditions of security. The topics addressed in this first phase were; *Globalisation and Security*, *Governance in Plural Societies and Security* and *Environment and Security*.

To conceptualize and develop the program a planning and coordinating conference was held at Rajendrapur in Dhaka in October 1999. Twenty leading intellectuals from all of South Asia came together to consider the various facets related to this emerging concept of human security and shape the project.

This book is a result of a two year-long study and workshops enquiring into the ways our neglect of the environment impacts on security in South Asia today. My sincere thanks to Dr Adil Najam, the Project Director, for his high intellectual input, personal dedication and leadership in this task which has made this project successful. Contributions have come from eminent intellectuals and experts from different countries of South Asia in what is truly a path breaking effort at generating greater understanding on this critical issue of our times.

DIPANKAR BANERJEE
Executive Director
Regional Centre for Strategic Studies
Colombo, Sri Lanka.
2002

Chapter One

Environment & Security: Exploring the Links

ADIL NAJAM

LIKE ALL CONCEPTS that are fundamental to the definition of their disciplines, the term 'security' has remained a subject of constant enquiry, debate and evolution. Although the extent of exactly *what* is being 'secured', *for whom*, *by whom* and *how* have long been in some contention, the interest in what are now called 'non-traditional' security issues gained considerable momentum in the wake of the end of the Cold War.

In the last half century the term 'security' was primarily a matter of states and their military alliances, principally applied to the 'security' of borders and institutions from outside threats (*see* Dalby 1992). The bi-polar nature of world dynamics that prevailed during the period intensified the emphasis on external threats. Such a definition, although considered minimalist by some, is accepted by many because military threats to security are easily identifiable and carry clear and often extreme consequences. In contrast, non-military threats within nations, such as poverty, social vulnerability or ecological resiliency are generally not perceived as concrete and tangible. A key conceptual difference between the two approaches is that the traditional definition of security presupposes that threats arising from outside the state are more dangerous to the state than threats that arise within it.

Recent debates on whether and how the concept of security might be expanded beyond issues of geo-polity, international power-balance, military strategy and statecraft have been both intense and rich (Galtung, 1982; Ullman, 1983; Mathews, 1989; Walt, 1991; Dalby, 1992; Buzan, 1991; Jones, 1995). One strand of this debate on non-traditional security issues relates to the increase in attention paid to the connection between environment and security. Scholarly discourse in this area has been prolific, though not always conclusive (Westing, 1988; Gleick, 1991; Libiszewski, 1992; Myers, 1993; Homer-Dixon, 1999; Deudney and Matthew, 1999; Dabelko, Lonergan and Matthew, 2000; Diehl and Gleditsch, 2001a). It is argued that environmental issues require new ways of thinking about policy, including security policy, because the environmental problematique is distinct by its very nature and must be kept separate from many other 'traditional' issues (Thomas, 1992; Alker et al., 1993; Elliott, 1998; Najam, 1994). The means of international environmental policymaking tend to be more consensual and cooperative than in other arenas (Stone, 1993; Susskind, 1994; Porter, et al., 2000; Najam 2000b), and these issues involve a very different set of actors, especially including nongovernmental organizations (Princen, et al., 1995; Najam, 1996, 1999 and 2000a; Mathews, 1997; Fisher, 1998).

The essays in this volume seek to engage in the above debates, and to do so from a particularly South Asian perspective. This is not an attempt to apply the frameworks posited in the larger available literature to specific cases from South Asia. Instead, this collection seeks to study how the link between environment and security manifests itself in the South Asian context as well as how this manifestation fits into, or negates, the thrust of the existing literature. While it is not the explicit goal of this volume to posit and defend a new and better framework for studying questions of environment and security, it is our hope that we will be able to contribute to such a framework. The immediate goal of this volume is more modest: to understand, in a policy-relevant way, the environment and security trends, threats, and opportunities for South Asia as a region and its component countries.

Environment and Security: Are the Links Real?

It has been argued that the environment is among the earliest and most pervasive sources of human conflict – and therefore of security concerns. Water, for example, has been – and remains – one of the most persistent sources of conflict at every level: international, national, community, and even individual. Energy is similarly one of the most potent motifs of environmental as well as security issues at every relevant level, from supranational to individual. Looking at environment and security links from a different perspective highlights the deep relationship between the deterioration of environmental quality – whether it is in the form of urban pollution, water contamination, soil degradation, deforestation, or biodiversity loss – and human well-being. Ultimately, a threat to human well-being can be seen as a threat to human security. Indeed, one could argue that the wrong end of a smoke stack can be as much of a security concern to humans as the barrel of a gun.

For many, the linkage is patently obvious and beyond debate. As Mark Halle (2000: 1; emphasis added) points out in his foreword to a literature review on the subject, "the relationship between environment and security *feels* right. It seems intuitively correct to assume a direct correlation between environmental degradation on the one hand and social disruption and conflict on the other."

While this intuitive feeling is widely shared, it is not universally accepted. On the one hand, there is now a sizeable literature that seeks to broaden the definition of the word 'security' in order to position environmental concerns as a central issue on the security agenda. Table 1.1 (based primarily on citations quoted in Diehl and Gleditsch, 2001b: 1-2) presents a sampling of views from those who consider the connection between environment and security to be overwhelming and obvious. On the other hand, however, many are not entirely convinced by the passion or the conviction with which such views are articulated. At the charitable end of this scale, Diehl and Gleditsch (2001: 2) criticize the literature on environment and security for offering "insights without evidence"; at the less sympathetic end, Daniel Deudney (1991) considers the zeal to establish a connection between environment and security an example of "muddled thinking."

Table 1.1
Environment and security: An 'obvious' connection?

WCED (1987: 290): *"Nations have often fought to assert or resist control over war materials, energy supplies, land, river basins, sea passages and other key environmental resources."*
Colinvaux (1980: 10): *"History has been a long progression of changing ways of life and changing population... [with] wars, trade and empire [the results]."*
McMichael (1993: 321): *"The end-stage of unequal power relations and economic exploitation in the world will be tension and struggle over life-sustaining resources. Fossil fuels, freshwater, farming and fish have already become the foci of armed struggles."*
Renner et al. (1991: 109): *"Throughout human history, but particularly since the system of sovereign states, struggles over access to and control over natural resources... have been a root cause of tension and conflict."*
Myers (1993 :21): *"National security is no longer about fighting forces and weaponry alone. It relates increasingly to watersheds, croplands, forests, genetic resources, climate, and other factors... that taken together deserve to be viewed as equally crucial to a nation's security as military prowess. The situation is epitomized by the leader who proclaims he will not permit one square meter of national territory to be ceded to a foreign invader, while allowing hundreds of square miles of topsoil to be eroded each year."*
Lodgaard (1992: 119): *"Where there is environmental degradation, or acute scarcity of vital resources, war may follow."*
Hassan, (1991: 64): *"Environmental deterioration leads to political deterioration, which engenders national and international insecurity."*

Brock (1991: 409): *"Control over natural resources has always been important in enabling a country to wage war."*

Ehrlich and Ehrlich (1970: 425): *"Population-related problems seem to be increasing the probability of triggering a thermonuclear Armageddon."*

Opschoor (1989: 137): *"Ecological stress and the consequences thereof may exacerbate tension within and between countries."*

Galtung (1982: 99): *"Wars are often over resources [and] destruction of the environment may lead to more wars over resources... environmental effects make a country more offensive because it is more vulnerable to attack and because it may wish to make up for the deficit by extending the ecocycles abroad, diluting and hiding the pollution, getting access to new resources."*

UN Security Council (1992): *"The absence of war and military conflicts amongst States does not itself ensure international peace and security. The non-military sources of instability in the economic, social, humanitarian and ecological fields have become threats to peace and security."*

In his review of the subject, Marvin Soroos (1994: 318-319) identifies four arguments that are commonly put forth by those who envision a strong link between environment and security:

- The ***conceptual*** argument that "security implies freedom or protection from serious threats to human well-being... [therefore] whatever poses such a threat, be it in military, economic, resource, food, or environmental realms, becomes a security problem." The works of Ullman (1983), Mathews (1989) and Myers (1993), for example, tilt heavily towards such arguments.
- The ***theoretical*** argument focuses principally on "empirical cause-and-effect relationships, in particular the potential of major environmental changes to generate and intensify [violent] conflict between and within states." This argument most identifies with the work of Thomas Homer-Dixon and his colleagues (see, Homer-Dixon, 1994 and 1999; Homer-Dixon and Blitt, 1998) but also with the research of Arthur Westing (1988), Peter Gleick (1991) and others.

- The ***political*** argument for linking environment and security seeks to "advance the environmental cause by taking advantage of the potency of the term security... [and] bestow the [environmental] problematique with a greater sense of urgency that elevates it to the realm of 'high politics'." Overtones of such a rationale are apparent in the writings of many who write on the subject from an environment-centric perspective.
- Finally, the ***normative*** case "presumes the primacy of environmental values and the threat that modern civilization poses to them" and seeks to place "societal values in a more appropriate hierarchy." One might cite authors such as Ehrlich and Ehrlich (1970), Brock (1991) and McMichael (1993) as exemplars of this argument.

While this characterization of the key facets of the argument remains quite valid, the discourse on environment and security has matured quite rapidly in recent years. Marc Levy (1995) labeled the first wave of research on environment and security as being preoccupied with rewriting the security agenda. Characteristic of this wave of research was Ullman's (1983: 133) attempt to redefine security as follows:

> A threat to national security is an action or sequence of events that: 1) threatens drastically and over a relatively brief period of time to degrade the quality of life for the inhabitants of a state, or 2) threatens significantly to narrow the range of policy choices available to the government of a state or to private, non-governmental entities (persons, groups, corporations) within the state.

Others, such as Marvin Soroos (1997: 236), have also tended to define 'security' very broadly, as "the assurance people have that they will continue to enjoy those things that are most important to their survival and well-being." The attempts to 'play' with the definition of security so as to insert environmental concerns were resented by a number of scholars of international affairs. They therefore found it necessary to call attention to the ambiguous nature of the link between environment and security (Deudney, 1990; Levy, 1995; Gleditsch, 2001).

The more recent literature on the subject, what Levy (1995) calls the 'second wave', has shifted away from this thrust in two important ways. The first is to move away from the notion of security and towards the term 'conflict'. This has a number of important implications, including a) bypassing the conceptual debates about the definition of 'security', b) getting a relatively more easily testable variable for analysis, and c) moving from an interstate to an intra-state focus. The second conscious shift made by the literature is toward more empirical research based on case studies in order to understand exactly *how* environment and conflict are linked. While there are multiple concentrations on this type of research in Europe and America, the most prominent work in this direction comes from the Peace and Conflict Studies Program at the University of Toronto. Eight key findings of this stream of research were highlighted in the Program's 1998 book *Ecoviolence* (Homer-Dixon and Blitt: 223-228):

- Under certain circumstances, scarcities of renewable resources produce civil violence and instability. The role of this environmental scarcity is often obscure, although it generates intermediate social effects that analysts often interpret as immediate causes of conflict.
- Environmental scarcity is caused by the degradation and depletion of renewable resources in addition to the increased demand for these resources and/or their unequal distribution. These causes often reinforce and interact with one another.
- Environmental scarcity often encourages powerful groups to capture valuable resources and prompts marginal groups to migrate to areas that are already ecologically sensitive.
- If social and economic adaptation is unsuccessful, environmental scarcity constrains economic development and contributes to migrations.
- In the absence of adaptation, environmental scarcity sharpens existing distinctions among social groups.

- In the absence of adaptation, environmental scarcity weakens governmental institutions and states.
- The above intermediate social effects of environmental scarcity can, in turn, cause ethnic conflicts, insurgencies, and coups d'état.
- Conflict generated in part by environmental scarcity can have significant indirect effects on the international community.

Such conclusions are characteristic of the second wave of research on 'environment and conflict'. While these results exemplify the ways in which this research has distanced itself from traditional 'security' debates, it has not been without its own problems. On the one hand, by shunning the language of security (except in the restricted terms of violent conflict) it stands open to the charge of retreating from the strong conceptual arguments an earlier generation of scholarship had made on why security *should* be redefined. On the other hand, many still fault it for lessening the importance of the issues due to an imprecise methodology and an ambiguous conceptualization.

In a particularly piercing review of this scholarship, Nils Petter Gleditsch (2001) identifies nine persistent problems with this new wave of research on environment and security. These include tendencies toward a) confusing resource scarcity with environmental degradation; b) polemically exaggerating the security aspect; c) overlooking important political, economic and cultural variables; d) devising complex and untestable models; e) using a biased selection of cases; f) reversing causality by inordinately focusing on environmental stress as a cause of conflict, g) using the future, rather than the past, as evidence; h) inappropriately generalizing between civil and interstate war; and i) muddling up different levels of analysis. In essence, such criticism constitutes a conceptual as well as a methodological challenge. In their response, Schwartz, et al. (2001) contend that Gleditsch seeks to impose a "methodological straightjacket" (p. 274) of "conventional, quasi-experimental methodology" (p. 273) which is inappropriate for research in this area because of its inherent complexity; they defend both their methodology and their conceptual framework and stress that "methodological pluralism" is necessary to make sense of a subject as complex as this (p. 291), as opposed to the traditional, unilateral approach that Gleditsch would have them take.

This debate is presented here to illustrate the vitality and the passion of the conceptual as well as methodological discourse in this area. Indeed, the faultlines are far more complex than a binary opposition between those who see a strong link between environment and security and those who don't. For example, Gleditsch (2001: 263) points out that, "Homer-Dixon (1991: 83) criticized earlier writing on [environment and security] as 'anecdotal' and has lambasted the work of a well-known environmentalist (Myers, 1993) as being 'marked by an almost complete absence of empirical rigor and theoretical structure' (Homer-Dixon, 1999: 184)." Ironically, Homer-Dixon's own case studies have been criticized by Marc Levy (1995: 56) for offering "more anecdotes, but not more understanding."

The point to be made, however, is that this area of research is still evolving; more importantly, it is an area characterized by rich and lively debate. There is certainly merit to the conceptual and methodological arguments of the critics (Deudney, 1990 and 1991; Levy, 1995; Gleditsch, 2001). Even as the debate continues to rage, however, there is nonetheless an emerging unison of interest between those coming from the security side of the question and those arriving from the environmental side. Indeed, the debate has now begun to also capture popular attention via the works of authors like Robert Kaplan (1994).

Moreover, the debate has now earned both scholarly and policy salience. From the scholarly perspective, there is (at the very least) the acknowledgement that this is an area worth studying in order to understand and explicate the *extent* of the linkage (implying that it does exist and the debate is on its strength rather than its validity). From the policy perspective, the language of environment and security has begun to seep not only into the documents of international organizations such as the United Nations and the North Atlantic Treaty Organization (NATO) but also into military establishments; indeed, the US Department of Defense now has a permanent position for a 'Deputy Under Secretary for Defense (Environmental Security).'

To summarize by way of responding to the question at the beginning of this section, there is now a growing consensus that the link between environment and security actually does exist. However, even though we have an improving understanding of the nature and extent of this link, it remains contested on both conceptual and methodological grounds. Moving forward from this, there is much interesting work being done both in terms of advancing the frameworks and ideas discussed above and in exploring related but new directions. The currently-emerging interest in looking at the issue from the perspective of human security and embedding it within the concept of sustainable development is of particular importance to this volume. The following section will outline key elements of this new development and propose a framework for organizing the various arguments before concluding with an outline of the chapters that follow.

Broadening the Base: Focusing on Human Security

The literature on environment and security has evolved over the years from its early focus on trying to expand the definition of 'security,' to the incorporation of environmental and related concerns, to its more recent preoccupation with understanding how environmental change can be a cause or amplifier of violent conflict. An emerging trend within this evolution has been a move toward greater emphasis on the concept of *human* security (Dabelko, Lonergan and Matthew, 2000; Elliott, 2001). The conceptual basis of most of the papers in this volume can be situated clearly on this emerging trend.

This broadening of the base is not in opposition to the earlier trends of redefining security or of mapping the environmental roots of violent conflict. In fact, it is an outgrowth of these trends. Indeed, many of the early attempts to broaden the definition of 'security' used very similar language to those who debate on 'human security' today. For example, consider the following definition from Norman Myers' *Ultimate Security* (1993: 31):

> … security applies most at the level of the individual citizen. It amounts to human well-being: not only protection from harm and injury but access to water, food, shelter, health, employment, and other basic requisites that are the due of every person on Earth. It is the collectivity of these citizen needs – overall safety and quality of life – that should figure prominently in the nation's view of security.

By a similar token, those who have focused on explicating the environmental causes of violent conflict have also brought the debate closer to the notion of human security, most noticeably by focusing on intra-state, and often local, insecurities. In sum, they have each pushed the debate towards "the concept of 'human security' [which] offers a third perspective that allows us to move beyond conventional security thinking, appreciates both the local and global dimensions of the many insecurities experienced by real individuals and groups, and identifies useful ways of linking security and development policies" (Dabelko, Lonergan and Matthew, 2000: 48).

While the concept of human security has earlier roots, its recent prominence comes from the 1994 *Human Development Report* (UNDP, 1994) of the United Nations Development Programme (UNDP). Suhrke (1999: 269) points out that "while offering an imprecise and controversial definition, the starting point for the UNDP was poverty rather than war – but 'security' suggested an escape from both." The currency of the concept was further advanced by the importance given to it in the report of the Commission on Global Governance (CGG, 1995). Both reports tried to shift the direction of the security discussion by focusing on issues of 'human life and human dignity' rather than on weapons and territory.

Lorraine Elliott (2001: 449) points out two dimensions of the human security paradigm that are of particular relevance:

> The first is that the concept of 'human security' provides an antidote to the more conventional focus on states, borders and territorial integrity. The answer to the question, 'security for whom' is not the state but the individual and communities, which suggests that even when a state is

> secure from external threats or internal instabilities, security for its people is not guaranteed. Protecting individuals and communities from the consequences of environmental decline (in this case) is therefore a security issue. The second dimension is that human insecurity (which includes equity, gender, human rights and identity concerns) is a central factor in social tensions and political instabilities and conflicts that can... become a feature of state insecurity.... If peoples and communities are insecure (economically, socially, politically, environmentally), state security can be fragile or uncertain. Environmental scarcity becomes a distributive equity problem rather than one simply of market failure, externalities or zero-sum calculations about access to resources and environmental services.

Indeed, as Elliott (2001: 449) recognizes, the human security paradigm "turns the conventional security aphorism – secure states means secure people – on its head." Dabelko, Lonergan and Matthew (2000: 48-49) add that the concept also "helps understand the complex interactions that determine the relative distribution of security and insecurity." They point out that "under certain conditions, such as war, the distribution and composition of force may be the most important determinant of security and insecurity." However, "in many other situations, security and insecurity will be most closely related to poverty or resource scarcity or social discrimination." Importantly, this leads to the conclusion that "in these cases, traditional security institutions may have only a minor contribution to make, or none at all." Indeed, most of the chapters in this book validate this finding.

While Dabelko, Lonergan and Matthew (2000) also point out the similarity between the goals of enhancing human security and sustainable development, they would agree with Astri Suhrke (1999) that a key relationship exists between the concepts of 'human security' and 'human development'. For the *Human Development Report* (UNDP, 1994) itself, "human development is a broader concept, defined as a process of widening the range of people's choices. Human security means that people can exercise these choices safely and freely." Suhrke (1999: 270-271) also argues that this relationship is more important to understanding the concept of human security:

> There are two possible starting points for exploring the substantive core of 'human security'. One is in relation to the security of states, the other in relation to human development.... The major contribution of the 1994 UNDP report was its attempt to define human security and human development, and sort out their relationship. The result, however, was confusingly circular. 'Human security' was presented both as an end-state of affairs – 'safety from such chronic threats as hunger, disease and repression' – and a process in the sense of 'protection from sudden and hurtful disruptions in the patterns of daily life'.... Human security was seen as essential for human development; without minimal stability and security in daily life, there could be no development – human or otherwise. But the obverse was true as well. Long-term development that improves social and economic life would produce human security, the UNDP report concluded. In this reasoning, there is no difference between development and human security, or between process and end-state.

Suhrke (1999: 271) goes on to argue that the critical distinction is "between (human) *development* and (human) *security*"; the first is about "long-term structural change" and the second about resilience to "sudden crisis-like disruptions". Most of our authors are concerned with human security in this more nuanced sense.

In trying to place this emerging interest in human security within the context of the evolution of the environment and security debate, one might propose a simple heuristic. In simplifying, for the purpose of exposition, Figure 1.1 conceives of an environment and security 'space' that is defined on one axis by the focus of analysis (ranging from state-centered to society-centered analysis) and on the other by sources of insecurity (ranging from violent conflict to social disruptions).

As we have already discussed, the early literature on the subject was concerned predominantly with state-centered discussions. While it did flirt with expanding the discussion of insecurity beyond traditional confines to also include social disruptions, it was mostly focused on inter-state conflict (since the audience of this literature was mostly restricted to the 'traditional security community'). Hence, the emphasis of the environment and security analysis very often turned to discussions of whether or not ***interstate war*** was a likely outcome. The 'second wave' of the literature honed in on the emphasis on violent conflict but made the focus of analysis more society-centered. The emphasis of the discussion, therefore, moved to whether and to what extent environmental change was a trigger for ***civil strife***. The new focus on ***human insecurity*** is also society-centered, but is more concerned with social disruptions (rather than violent conflict) as the principal source of insecurity. One of the key benefits of using such a heuristic is that it begins to point us towards other formalizations of the problematique that are not yet dominant in the available literature. In this case, Figure 1.1 points out the insecurity that emerges from social disruptions at the level of the state rather than the level of society. It is proposed that such insecurity is most likely to manifest itself as ***institutional failure*** and be best understood through a focus on the mechanisms of societal governance.

Figure 1.1
Organizing the environment and security discussion

		Sources of Insecurity	
		Violent Conflict	*Social Disruptions*
Focus of Analysis	*State-Centered*	**Interstate War**	**Institutional Failure**
	Society-Centered	**Civil Strife**	**Human Insecurity**

Although this heuristic is still tentative and in need of empirical validation, it provides us with one way to organize and understand the discussion. Importantly, we now have a means to place the discussion in this volume within the context of the literature reviewed. The chapters in this volume lie very much in the right-hand half of Figure 1.1, and predominantly in the bottom-right quarter. The emphasis is very much on environment-related insecurities as manifest in social disruption rather than in outright conflict. While the focus of these chapters is predominantly society-centered, it is important to note, as discussed in Chapter 10, that the state-centered aspects of such disruptions do tend to highlight failures of governance and institutions.

South Asia in Context

As already noted, the context of our subsequent analysis is the South Asian subcontinent. Detailed descriptions of the five largest countries of the region are provided in the next five chapters. They deal individually with the environment and security nexus in India, Pakistan, Bangladesh, Sri Lanka and Nepal, respectively. While it is not our purpose to repeat any of the details in this introductory chapter, it is useful to add a snapshot of South Asia as a region. We define the region here by membership in the South Asian Association for Regional Cooperation (SAARC), which was formed in 1987 and includes Bangladesh, Bhutan, India, Maldives, Nepal, Pakistan and Sri Lanka.

Home to nearly a fourth of all humanity, the South Asian subcontinent is a region where histories, geographies and politics are truly intertwined. The roots of the region's distinct identity predate the histories of any of the countries that now constitute it. The region has been home to great empires of its own as well as prey to the lust of others. It is a region that has been familiar with insecurities of all kinds throughout its rich and tortured history; this still holds true today. The 1997 *Human Development in South Asia* Report (Mahbub ul Haq, 1997) described South Asia as 'the most deprived region' in the world. Certain elements of the rather depressing picture of the region that the report painted are worth repeating here:

- South Asia is the **world's poorest region** with a per capita GNP below even that of sub-Saharan Africa, which is home to 40 per cent of the world's poor and to over 500 million people below the absolute poverty line.

Table 1.2
South Asia's many roots of insecurity

	Bangla-desh	India	Nepal	Pakistan
Area in thousand square kilometers (1999)	144	3,288	147	796
Population Millions (1999)	128	998	23	135
Population growth Rate Annual (1990-99)	1.6	1.8	2.4	2.5
GNP per capita US Dollars (1999)	370	450	220	470
GNP per capita PPP Dollars (1999)	1,475	2,144	1,219	1,757
Under-5 Mortality Per 1,000 (1998)	96	83	107	120
Life Expectancy **M/F** (1998)	58/59	62/64	58/58	61/63
Adult Illiteracy % population 15 and above; M/F (1998)	49/71	33/57	43/78	42/71
Urban Sanitation % with access (1990-96)	77	46	34	53
Children in Labor (10-14) % of age group (1999)	29	13	43	16
Gini Index (1995-97)	33.6	37.8	36.7	31.2
Spending on Education % of GNP (1997)	2.2	3.2	3.2	2.7
Spending on Health % of GDP (1990-98)	1.6	0.6	1.3	0.9
External Debt % of GNP (1998)	22	20	31	41

Source: World Bank (2000)

Table 1.2 *(continued)*
South Asia's many roots of insecurity

Sri Lanka	Developing countries	World	
66	*101,487*	*133,572*	**Area** in thousand square kilometers (1999)
19	*5,084*	*5.975*	**Population** Millions (1999)
1.2	*1.9*	*1.7*	**Population growth Rate** Annual (1990-99)
820	*1,240*	*4,890*	**GNP per capita** US Dollars (1999)
3,056	*3,410*	*6,490*	**GNP per capita** PPP Dollars (1999)
18	*79*	*75*	**Under-5 Mortality** Per 1,000 (1998)
71/76	*63/67*	*65/69*	**Life Expectancy M/F** (1998)
6/12	*18/33*	*18/32*	**Adult Illiteracy** % population 15 and above; M/F (1998)
33	-	-	**Urban Sanitation** % with access (1990-96)
2	*13*	*12*	**Children in Labor** (10-14) % of age group (1999)
34.4	-	-	**Gini Index** (1995-97)
3.4	*4.1*	*4.8*	**Spending on Education** % of GNP (1997)
1.4	*1.9*	*2.5*	**Spending on Health** % of GDP (1990-98)
41	-	-	**External Debt** % of GNP (1998)

Source: World Bank (2000)

- South Asia is the **world's most illiterate region** and home to nearly half of all the illiterates in the world; there are more children out of school in this region than in the rest of the world combined; two-thirds of this wasted generation are female.
- South Asia is the **region with the highest human deprivation:** 260 million people lack access to basic health facilities, 337 million are without safe drinking water, 830 million are without rudimentary sanitation, and 400 million people go hungry every day.

South Asia may also be among the most militarized regions in the world. India and Pakistan have both declared themselves nuclear powers. They have fought three full-scale wars with one another and continue to have near-constant skirmishes on their borders, especially over the disputed region of Kashmir. Neither India nor Pakistan can afford their massive military expenditures: US$15 billion and US$3.5 billion, respectively. Indeed, as the *Human Development in South Asia Report* (Mahbub ul Haq, 1997: 80) pointed out, "India was ranked by the World Bank at 142 in terms of real per capita income (PPP dollars) between 1988 and 1992, yet it ranks first in the world in total arms imports. Pakistan's position is marginally better, at 119 in per capita income and tenth in arms imports. When most basic social services are missing in both India and Pakistan ... the rising defense burdens in these countries continue to impose prohibitive social and economic costs on their people." The other countries of the region, while nowhere near as committed to large militaries, are also burdened by military expenditures greater than they can afford, often because of internal threats.

Table 1.2 presents a brief profile of the five largest countries of the region. It is clear from the table that on all the variables presented, South Asian countries are not only significantly behind the world as a whole, but also well behind developing countries as a group (measured here as the average of all low and medium income countries). These, as we shall see in later chapters, are the roots of human insecurity in the region and end up having significant implications for the environment. Table 1.2 also highlights the fact that although there are important differences within the region (for example, in terms of education), there is a certain uniformity in the development profiles of the region. In essence, the table helps us make two important points about why South Asia should be the subject of a study such as this. First, this is very much a region that *can* be studied as a region; not only in terms of its historical legacy, but also in terms of its current developmental predicament. Second, this is a region that *should* be studied; the region's acute developmental depravations point towards the potential for equally acute, even violent, human insecurity in the future.

In order to get a more nuanced view of the environment and security concerns in the region, we first look at individual countries within it, and then turn our attention to selected issues and how they play out across the entire region. This volume is organized into two distinct sections. The next five chapters are each devoted to an in-depth study of environment and security issues in five countries. The essays on India, Pakistan, Bangladesh, Sri Lanka and Nepal each present a general picture of the environment and security concerns for that particular country as well as discussions of selected case studies within the given framework. Thematic chapters follow – one each on energy, land and water – which focus on three critical environment and security issues for the region as a whole. The last chapter in the volume highlights a set of five key crosscutting lessons that emerge from looking at all eight chapters together.

In Chapter 2, focusing on India, Vandana Asthana and Ashok Shukla provide a detailed review of the environment in the region's largest country and investigate the nature of links between security and environmental stress both domestically and internationally. They find, like many of the chapters in this volume, that environmental stress can and does translate to human insecurity, but it does not necessarily take the form of violent conflict. In looking at the region as a whole, they cautiously conclude that sustainable development has the potential to become a new paradigm for domestic and regional security. The view from Pakistan is less sanguine.

The focus of Chapter 3, by Shaheen Rafi Khan, is also sustainable development. His concern, however, is more about how the absence of sustainable human development can lead to environmental degradation, thereby triggering social inequity and injustice. The chapter uses two case studies to trace how environmental stresses can lead to human insecurity and conflict.

In Chapter 4, Atiq Rahman, Zahid Chowdhury and Ahsan Ahmad present a detailed review of the stresses placed on Bangladesh's environment. From this they draw the links to human security issues and conclude that poverty is, and is likely to remain, the most important source of environmental vulnerability and insecurity in Bangladesh.

In studying Sri Lanka, Chapter 5, Sarath Kotagama takes a rather different tack, presenting a series of short case studies which highlight the various ways in which security concerns can come to the fore in what might otherwise be considered 'straightforward' episodes of environmental stress. His conclusion, however, echoes the findings of other chapters, particularly in relation to his claim that the environment-security nexus is integrally tied to the cultural and institutional robustness of societies.

In the last of the country-chapters, Ajaya Dixit and Dipak Gyawali (Chapter 6) apply a risk-based, cultural theory perspective to a water-sector case study with domestic as well as regional livelihood implications. Their analysis concludes that an atmosphere of insecurity can be created when environmental stresses interact with societal vulnerability, disruptive development, and perverse markets; it is this volatile mix, rather than the environmental stress alone, that can become a trigger for insecurity or violence.

The following three chapters concentrate on specific themes and study them across South Asia. In Chapter 7, which looks at energy issues, Kumudu Gunasekera and I adopt a multiple frameworks analysis – using indicators of energy efficiency, environmental impact, and external dependence – to plot the energy security profile of the South Asian countries separately as well as that of the region as a whole. We conclude, not unlike many of the preceding chapters, that structures of governance and policy – rather than natural endowments alone – are the key to enhanced energy security.

A similar conclusion is reached in Chapter 8, which looks at land and food security, even though it adopts a very different analytical methodology. In this chapter, Khalid Saeed uses a system analytic model to review the food security profile of selected countries in the Asia-Pacific region, with a particular focus on South Asian countries. As in the chapter on energy, the conclusion is reached that institutional design is the most critical determinant of land-related environmental insecurity, particularly in terms of food security.

The last of the thematic chapters looks at the links between water and insecurity. In this chapter, Ramaswamy Iyer reviews the history of interstate water relations in South Asia. In doing so, he questions the very utility of the term 'security' in the environmental context and concludes that there is little likelihood of interstate conflict in South Asia on issues related to water. However, he adds that there is much room for improved regional cooperation in this area.

Finally, Chapter 10 reviews the findings of the chapters that precede it in order to draw out the key conclusions and lessons. It suggests that the dominant common strand between all the chapters is that the defining, even causal, link between environment and security in South Asia is poverty. In addition, it argues that human security itself – more than just violent conflict – is the lens through which this link is best understood in the South Asian context. More specifically, Chapter 10 outlines the following five lessons that emerge from this volume as a whole:

- For South Asia in particular, and developing countries in general, environment and security are best conceptualized within the context of sustainable development.
- The challenge of environment and security in South Asia is principally a challenge at the domestic, even local, level; but it is a challenge common to the region.
- The challenge of environment and security in South Asia is, at its core, not just a problem of resource endowments or geography but, quite distinctly, a problem of institutions and governance; it is only because the issue is the latter rather than the former that we have the ability to change the situation.
- The possibility of an eruption of interstate violence in South Asia over environmental issues is slim; however, given the region's history of distrust and dispute between nations, environmental differences can add to existing tensions and

apprehensions, thereby perpetuating the general sense of insecurity that pervades interstate relations in the region.

- There is the potential – albeit small – for a new generation of security relations in South Asia, based on the principles of mutual trust, harmony and cooperation rather than on legacies of distrust and dispute, to emerge there around the nexus of environment and security.

About This Book

This book is the result of a larger project on 'Non-Traditional Security Issues in South Asia.' The project, housed at the Regional Centre for Strategic Studies (RCSS), Colombo, Sri Lanka, was supported by the Ford Foundation and was itself part of a larger Asia-wide research initiative on the same subject. The focus on this particular component of the project (the other two being *Globalization and Security* and *Governance in Plural Societies and Security*) is on the relationship between environment and security in South Asia.

South Asia (and the nations within it) has already been the subject of earlier research on environment and security (Myers, 1989 and 1993; Hassan 1991 and 1992; Islam, 1994; Gizewski and Homer-Dixon, 1998). The purpose of this project, then, was to advance this research with a decidedly policy-relevant focus, to build institutional capacity and expertise within the region for research on this topic, and to facilitate improved networking between individuals and institutions focusing on issues of this nature. This volume represents one step in this process.

A conscious attempt was made not to seek, or impose, a particular view of what is meant by the 'environment and security nexus'. Indeed, different authors have come up with very different (and sometimes opposing) viewpoints of what this term means. The project at large, and this volume, have sought to embrace and elicit this kind of diversity so that the issues may be covered from different conceptual directions. A diversity of viewpoints is, in fact, the most prominent feature of this volume. It represents not only the authors' different ideas on the subject, but also their varied interdisciplinary backgrounds.

The authors all hail from South Asia; they bring with them a richness and an authenticity of experience. The writing team for this volume included three authors each from India, Pakistan and Bangladesh, and two each from Nepal and Sri Lanka. Importantly, the writing team included scholars with expertise in varied disciplines including political science, environmental engineering, resource economics, geography, wildlife ecology, system dynamics, hydrology, urban planning, and anthropology, to name a few. This disciplinary and experiential diversity has allowed us to view the environment and security problematique through multiple lenses, all from a shared South Asian perspective.

Acknowledgement

As the Project Director and editor for this volume, I would like to express my deep gratitude to Dipankar Banerjee, Executive Director of the Regional Centre for Strategic Studies (RCSS), without whose constant guidance and encouragement this effort would have neither commenced nor concluded. I have learned a tremendous amount, personally and professionally, under his gracious mentorship.

My gratitude also goes out to all the authors who were so generously forthcoming with their time and intellect. The intellectual diversity of the writing team, as well as the common rootedness of all the authors to South Asia provided a unique and enriching learning environment.

Finally, I would like to thank Kumudu Gunasekera who not only co-authored a chapter with me but also provided general research assistance for the project, and to Emily Thompson, Claire Norris, and Ramya Ramanathan for their editorial assistance.

Chapter Two

Sustainable Development as a New Security Paradigm for India

VANDANA ASTHANA
ASHOK C. SHUKLA

SOUTH ASIA IS at a defining moment in history. Globalization has ushered in the liberalization of economies and has been instrumental in enhancing interdependence among nation states. Environmental regimes are in the making to realize sustainable development and environmental amelioration. Expedient exhaustion of and threats to natural resources can never cease under exploitative regimes. Environmental challenges are not only causing stress within states but could potentially forge conflicts, micro- or macro-wars, between states. Militarization apart, resource depletion, degradation, and alienation, as well as increased demand to obtain and corner scarce resources hold potential causes for conflict between competing users in

South Asia and are likely to make the region more volatile and insecure in the future.

As a vast nation with its massive size, natural resource base, agricultural and industrial productivity, biodiversity, and enormous human resources, as well as in a geopolitical and cultural sense, India stands distinguished among the nations in South Asia. While there has been a growing global recognition of this vast resource and its potential, simultaneously there also exists evidence to undermine it in the long term; not least from development strategies that might erode the nation's natural resource base and expose it to strife and insecurity.

The nature of ecological threats to the physical security of India also has transnational implications. While the potential of ecology-related threats of conflict is episodic, the gravity of these threats in the context of resource depletion and the sharing of river waters as well as the consequences of acid rain, over-fishing, and other environmental stresses, could endanger security within the country and in interstate relations. The ecology-security nexus presents new security challenges for India and for South Asia. At the national level, this link points in the direction of environmental scarcities, poverty, and civil strife weakening the state's legitimacy; on the regional level, it calls for enhanced understanding of the region as an integrated set of ecosystems.

As the various chapters in this volume highlight, there are myriad causes that could trigger conflicts within states over environmental issues and resource scarcity. Homer-Dixon (1994) identifies three types of environmental scarcities that exist and interact simultaneously: demand-induced scarcities caused by population growth and increase in per capita consumption; supply-induced scarcities causing degradation and depletion of resource; and structural scarcities arising from unequal social distribution, thereby concentrating resources in the hands of a few individuals and causing social segmentation. Thus, environmental stress is an important part of the web of casualty associated with any conflict and 'straightforward' environmental issues are often deeply interlinked with challenges to national security. Indeed, the concept of security is undergoing a change, from an exclusive focus on territorial security to a much greater emphasis on security of people -- from security through arms to security through sustainable development.

There is now a growing literature dealing with the links between environment and security (Dalby, 1992; Ullman, 1983; Myers, 1993; Mathews, 1989; Homer-Dixon, 1991, 1994; Vandana, 1992; 1999; Shukla and Vandana, 1999; Mollinga, 2000; etc.). Scholars of various persuasions have battled with competing views and notions of these links and posited different ideas and ideals. Among these, a particular place of prominence is given to the idea of human security. However, the paradigm of human security signals a divergence from conventional notions of national security. Human security embodies individual well-being through human development, food sufficiency, employment and environmental quality. Fundamentally, human security encompasses the security of people against threats to life, health, dignity, and personal safety. The concept of security has broadened into a multidimensional approach and a holistic view is emerging. There seems to be no line drawn between 'environmental issues in a security paradigm' and 'environmental security' itself, because both impinge upon each other and are interconnected and interlinked.

Environmental change is conceived in terms of nature and the extent of environmental stress. Environmental resource scarcity and degradation cause environmental stress. The consequences of environmental stress include poverty, food insecurity, poor health conditions, displacement, and disruption of social and political institutions, which then contribute to conflict under unfavorable conditions. These factors comprise economic vulnerability and resource dependency, institutional, socio-economic and technological capacity, cultural and ethno-political factors, patterns of perception, the potential for violence, internal security structures, participation in political stability, international interaction, and mechanisms of conflict resolution (Lietzmann and Vest, 1999). These, in turn, influence environmental change, causing social, economic, political and demographic consequences, which in turn have an impact on security.

The goal of this chapter is to understand these interconnections by using India as a case study. India is a country striving to carve a pivotal role in shaping the destiny of environment and security concerns in

South Asia. What are the promises and pitfalls of India in shaping South Asian security? Where does India stand in South Asia in terms of resources? What are the issues concerned with peace and security? What are the existing and potential threats in the ecology-security nexus of India and South Asia? What are some alternative strategies India could employ in order to save itself and the region from getting trapped in the catastrophic whirlpool?

In searching for answers, we begin first by briefly tracing India's long, rich environmental history. This is followed by stage-setting sections describing the context of this study – the state of India's environment today. The remainder of the paper will investigate the links between environmental and security issues in terms of their implications for domestic strife and regional friction. Finally, we explore the possibility of a sustainable development-based security paradigm becoming the basis of lasting security, not only for India but for the South Asian region as a whole.

History of Indian Environmentalism

Environmentalism as a basic ideology is an explicit element of the history of India from time immemorial and has had profound impacts on Indian scripture, religion and art. Living in mutual harmony with nature has been a part of India's cultural as well as survival heritage and has manifested itself in a rich legacy of cultural and ecological diversity. This diversity is manifest even today in the existence of stone-age hunters and gatherers like the colonists of Kerala, slash and burn agriculturists like Maria Gonds of Central India, nomadic pastoralists like the Bhotias of the Himalayas and the Bishnois of Rajasthan, subsistence agriculturists of the dry Deccans, commercial agriculturists of the Sutlej and Godavari deltas, as well as the advanced industrial urban centers like Bombay and Calcutta (Karve, 1961). India's ecological diversity is equally breathtaking, ranging from the sub-zero temperatures of the Ladakh plateau to the hot rainforests of Andaman and Kerala, and from the rainiest place on earth at Cherapunji, Meghalaya to the scorching desert of the Rann of Kutch.

Besides this physically diverse landscape, India is home to more than 45,000 different species, more than any other landmass of comparable size in the world.

To survive in, and with, an environment of such ecological and cultural diversity, traditional Indian societies evolved a prudent system of resource use, which was entrenched in a deep-rooted ethic of conservation and sustainable resource management. Many religious and folk traditions of religious and folk beliefs stressing the respect for nature and its resources are found in the tribal societies of northeastern and central India, where sacred forest groves are worshipped even today (Gadgil and Vartak, 1976). The hunter-gatherer societies gave way to cultivators, beginning in what is now Pakistan and northwestern parts of current India as early as 6000 B.C.. The Indus Valley civilization began around 2000 B.C., as recorded in relics found in Mohenjodaro (in modern-day Sindh, Pakistan). This civilization worshipped trees as givers of wisdom and life. Tree worship was performed in two ways. either the tree was personified and endowed with human attributes or it was worshipped in its material form. Animal worship was also a common practice in the Harappan period.

Indeed, ancient Hindu texts are a rich source of information and insight on the historical roots of Indian environmentalism. The Vedas are the oldest literature of the Indo-Aryans and one of the earliest Sanskrit texts. The *Rigveda* is the most important of these; it contains invocations addressed to various gods and goddesses relating to different personified powers of nature. The *Atharveda* is especially important for the history of science in India. It describes the magical properties of several plants. The *Yajurveda* lays down rituals and methods pertaining to performance of *yajnas* (sacrifices). The *Mahabaharta* is the bulkiest Indian epic; it describes not only the war between the Kauravas and the Pandavas but also contains the entire Bhagavad Gita, the discourse given by Lord Krishna to Arjuna during the war. It is a philosophy of action, destruction and creation.

The earth, according to the Vedas, is not only for human beings to enjoy but also for all other bipeds and quadrupeds, birds and insects and other creatures (Atharveda, 12.1.15). The most important aspect of Hindu theology is the belief that the Supreme Being actually gets incarnated in the form of various species (Srimad Bhagavatam, Book 1, Discourse – III: 5). Indeed, ancient Hindu societies understood God and *Prakriti* (nature) to be one and the same (Mahabharata, moksa, 182). It was the original human instinct to survive that compelled them to cooperate with, and benefit from, their environment. The Hindu concept of 'nature as divine' began in the hunter-gatherer stage and continued into the pastoral and agricultural societies. While these societies indulged in the clearing of forests and killing of wild animals, they did manage to retain an awareness of the need for ecological prudence and preserved many practices of the earlier tribal period. The importance of cattle in Vedic India was not merely economic but also ritualistic and symbolic. The cow had religious sanctity; cows had been worshipped since the *Rigveda* and are thought to be goddesses. An important aspect of animal worship among Hindus has been related to the incarnation of God in the form of fish *(Matsya Avatara),* tortoises *(Kachap Avatara),* boars *(Varaha Avtara),* dwarves *(Bamana Avatara)*, and man-lions *(Narsingh Avatara). Hanuman*, the ancestor of monkeys, rendered great service to *Bhagwan Ram* and because of this, monkeys are still shown reverence today. Many Hindu gods and goddesses use animals and birds as mounts.

Apart from the Vedic and Upanishad periods, Buddhism and Jainism further emphasized and strengthened the protection of animals, birds, and vegetation in a practice that was continued by many kings. Emperor Ashoka (273-236 B.C.) undertook various steps to safeguard the environment through public proclamations and plantations. Pillar edicts expressed concern over animals, plants, and trees and prescribed punishment for the murder of beings including ants, squirrels, and rats. Multiple uses of different trees and plants have been known since ancient times and their use in religious functions, marriages, and ceremonies are still popular today. Tree worship can be traced back to the prehistoric period as the earliest and the most prevalent form of religion. Nature in the ancient Indian philosophy was essential for life of human beings, and some of the trees and plants were abodes of gods and goddesses. *Rigveda* (10.97) regards plants as having divine powers and the epics and the *puranas* expanded this concept to include trees as having life and feelings of happiness and sorrow *(Mahabharata 18.15-*

17; Puranas, literally meaning 'old', deal with creation, genealogies of Gods, patriarchs, historical and mythological events of ancient times). Trees like *Tulsi, Ashoka, Neem, Pipal, Mahua, Semal, Kadamba, Bela, Soma*, *Mango* and others are associated with various gods and goddesses in the Hindu religion. The practice of large-scale tree planting in India is at least fifteen hundred years old. The *Matsya Purana* (59.159) and *Varaha Purana* (172-39) celebrate the tree plantation ceremony called *vanamahotsava* and state that those who plant trees will not go to hell. Cutting trees was considered a sin and a punishable offence (*Padma Purana*, 56.40.41).

The Brahminical era, which emphasized cultivation and the sacrifice of animals, led to strong protest. This protest found its expression in two new religions -- Buddhism and Jainism -- in the sixth century B.C.. The need to respect all forms of life, *Ahimsa* and nonviolence had considerable influence for over a thousand years in India. Even *Arthashashtra* -- Kautilya's manual on statecraft -- emphasizes the conservation of nature and prescribes various punishments for destroying trees and plants (*Shamashastra*, 1967). The *Mahabharata, Rajadharma nusasanparva*, (16.811), *Visnusmriti* (23.3846), *Charaka Samhita, Vimansthan* (3.2), *Rigveda* (7.49.2*)*, and *Vishnu Purana* (3.11.11-12*)* all mention the various aspects of environmental pollution pertaining to air, water, habitat, sanitation, etc., and prescribe certain rules and provisions for punishment in order to preserve and protect the environment. Thus, the culture of ancient India was based on a deep reverence, respect, and love for nature; the people worked towards the goal of human-nature harmony.

The influx of outside cultures and invasions added new cultural systems to the mix, displacing these ethical and moral values. However, even when nature was not revered in the religious sense, it remained a mainstay of India's cultural traditions. For example, the Mughal emperors Babur, Humayun, Akbar, and Aurangzeb all expressed considerable love for nature and backed it up with concrete policies and regulatory edicts that had clear environmental goals. These included rules relating to the protection of forests, habitats, species, and bodies of water. The emperor Jahangir was a born naturalist and his observations continue to inspire naturalists even today. The British invasion added western concepts, including naturalistic and preservationist tendencies.

By the mid-twentieth century, however, demographic, technological, economic, and social challenges had clearly tilted the balance away from nature and given birth to the first inklings of an indigenous environmental movement in India. A fresh perspective on India's ecological history and history of environmental movements can be found in the works of eminent historian ecologist Ramchandra Guha. Rabindranath Tagore and Mohandas Gandhi, both of whom had immense influence on India in terms of social and political ethics, were also great proponents of living in harmony with nature and may be considered among India's early modern environmentalists. Tagore's vision was based on an amalgamation of Indian tradition and local environmental needs. He propagated the view that a cooperative and considerate attitude towards nature would prepare people to shoulder the responsibility of an independent India. At the same time, Gandhi was advocating a 'close to nature' philosophy and preaching a simple life in co-existence with nature. Indeed, one might argue that Tagore and Gandhi were laying down the base for what would later become known as sustainable development. Their views laid the foundation of what has now become a thriving and diverse environmental movement.

People and Politics

The purpose of this section, and the next, is to provide the backdrop and context for our study. We begin here by briefly outlining the key issues in terms of India's contemporary geopolitics, institutions, and demography. This will be followed in the next section by a brief overview of the current environmental situation.

Geopolitics

India is the largest and most populous country in South Asia; it makes up four-fifths of the SAARC (South Asian Association for Regional Cooperation) region. India covers an area of 3,287,263 square kilometers. It is the seventh largest country in the world and unique in many ways from the rest of Asia. The spine of the great Himalayan Mountains in the north stretches southwards and, at the Tropic of Cancer, tapers into the Indian Ocean between the Bay of Bengal on the east and Arabian Sea on the west. India has a land frontier of about 15,200 kilometers, and the total length of its coastline (including the Lakshadeep, Andaman and Nicobar Islands) is 75,166 kilometers. India shares common borders with Afghanistan and Pakistan to the northwest, China and Nepal to the north, and Myanmar and Bangladesh to the east.

Historically, South Asian nations are linked by ties of race, culture, and religion, upon which they have built trade structures and political relationships. Geographically, they are proximate states sharing strong links as an aftermath of the subcontinent's partition in the 1940's. India, by virtue of its geographical area, economic and military developments occupies the central position in the region. Indian society is vast and heterogeneous; its many religions, ethnicities, and cultural traditions have much in common with the religions, ethnicities, and cultural traditions of its neighbors. Intermeshing people, culture, and religion imparts the potential for affinity for interstate relations in South Asia (Mansingh, 1982). In terms of socio-cultural links, a number of common languages are spoken or understood in more than one country. With the exception of India, which remains constitutionally secular, every other nation has adopted Hinduism, Islam or Buddhism as a state religion. However, one can often find a sizeable section of population belonging to other faiths and religions. The ethnic and cultural links between India and its neighbors often blur the distinction between foreign and domestic politics, and interstate relations tend to be under constant strain. India is not insulated from her neighbors or national identity, and fears of Indian hegemony are a sensitive matter to her neighbors. Fear of cultural domination exists more visibly in states that have been once part of India and in states where the institution of parliamentary democracy could not be established, as it has been in India.

The South Asian region has built-in problems of land frontiers, water-sharing, ethnic and cultural contiguities and India's relations in South Asia necessarily involve an intermeshing of concerns affecting political relations of these states, not just with India but also with extra-regional powers. It has been argued that the involvement of extra-regional powers is unlikely to build cooperative development, but if the states of the region recognize India's territorial integrity, then it seems possible to predict constructive and mutually compatible patterns of cooperative relations between India and her neighbors.

Specific problems relating to geopolitics have arisen among contiguous and proximate states in terms of territorial and maritime boundaries. The cause of the largest number of wars with India has been its geographical proximity with Pakistan; Kashmir has remained a bone of contention between the two states. Siachen is another such example. Other conflicts due to geographical proximity include the problem of foreigners in Assam and the immigration of a large refugee population from Bangladesh to India, which has unbalanced the northeast region (Perceival and Homer-Dixon, 1995). The geopolitics of shared water have been evident in the form of the Indus river conflict, which led to the Ravi-Beas Agreement between India and Pakistan, the Farakka Barrage in Bangladesh, the Wular Barrage in India, and the politics of the Indian Ocean region including fishing rights. Aside from traditional conflicts, these environmental stresses and degradations serve to increase the probability of future conflicts and deepen traditional differences.

Governance and Political Institutions in India

India is one of the largest democracies in the world. While democratic governance has its own strengths and weaknesses, perhaps the most significant accomplishment of India's polity is that despite some problems and the social complexities of linguistic, ethnic, class, caste, and sectarian differences, the democratic process has continued without interruption. Democracy is distinguished from other forms of governance in the region because it contains an internal mechanism of checks and balances on the abuse of power, ushering in gradual and peaceful transitions. The governing values of electoral democracy, cooperative federalism, secularism, and independent and nonaligned foreign policy are enshrined in India's constitution. A well-developed party system governs the national political system. Regional parties, coalitions, and independents have also gained momentum and are a regular feature in the nation's governance.

In recent years, however, this semblance of unity, compassion, and tolerance has been disintegrating, hampering inter-community relations, increasing fundamentalism, and ushering in the emergence of sub-national identities with deepening impacts on the rule of current governing institutions. As is common in many developing countries, political institutions in the region are under severe strain. For example, the criminalization of politics that characterizes India is not unlike the situations in neighboring Bangladesh or Pakistan. This criminalization refers to the large-scale, nearly institutionalized corruption that has become a permanent feature of Indian politics; for example, known and sometimes convicted criminals can and have become members of parliament. On the one hand, there is a steady decline in political ideology and morality; on the other, religious and ethical considerations among the political leadership are on the increase. Sycophancy and touts have flourished at all levels. Democracy is confined to the polls, and there is little real participation of people in governance after the elections. Survival in power has become the key principle. Judicial activism has increased in areas traditionally controlled by executives and legislature because of irresponsible behavior. Traditional power blocs joined by other economically upper-class communities control and share power even at the state level. In short, while democracy has been maintained against all odds, it does not always work as well as one would wish it to.

India has followed a developmental model that equates development with economic growth alone. This consumption-directed, supply-based approach tends to ignore or underplay the environmental impacts and concepts of sustainability. One of the most important effects of this developmental paradigm is environmental degradation and imbalances in the ecological life support system. This often results in conflicts between the state, environmentalists, and indigenous populations.

Population

The environmental problems faced by India are, to a large extent, a product of its burgeoning population, which in turn shapes society's perceptions of security. Demand and supply ratios also respond to the degree of population displacements and migrations. While mobilization and use of natural resources are important factors in the environment-development debate, demographic factors relating to population size, growth rate, density, and rural and urban distribution form important variables impinging upon environment and security issues.

Indian has a population of over one billion. Growing at a rate of 1.9 percent each year, the population is estimated to reach 1.16 billion by 2010 (Hassan, 1988). India's urban population is the largest in the world, despite the fact that 74.27 percent of our people live in rural areas. The metropolis of Mumbai is India's most populated city, with a total urban population of 12.6 million followed by Calcutta (11.0 million), Delhi (8.42 million) and Chennai (5.4 million) (Ministry of Information and Broadcasting, 2000). The rate of urban growth has since increased to over 4 percent. At this rate, half of India will live in urban areas by 2025 (Sinha, 1998).

This rapid increase in population is a threat to national development as it negates economic, social, and political achievements. While India has made major strides in economic growth, self-reliance, and modernization, these achievements fail to make an appreciable impact on the life of the common Indian. About 50 percent of its people still live below the poverty line and, on the basis of per capita income, India ranks 106 in the world. Rising subsistence demands have affected the traditional balance between the Indian population and resource availability. In order to survive, depleted resources are exploited beyond their regenerative capacity. The gains in agriculture and industry seem to be wiped out by population growth. India produces just enough food today to feed its population, but its food security -- and thus its political stability -- may be severely challenged as the population keeps on growing. (Lama, 1997)

There is also a correlation between population growth, poverty, and environmental degradation. Poverty is another name for social and economic deprivation. Lack of subsistence generates feelings of insecurity, compelling a life of drudgery, misery, and crime. Another predominant factor is the impact of population growth on health and quality of life. Roughly 40 percent of India's people suffer from malnutrition, high rates of infant mortality, and limited sanitation. Escalating social violence has continued to distract from the social and ecological agendas. In a setup of intense social stratification, new wealthy communities join the traditional power blocs – feudal landlords, upper castes, religious elements, etc. -- in continuing the cycle of subjugation. As the population increases, the problems of migration towards urban areas also increase. Conflicts over jobs, access to urban facilities, resource pressures on agriculture and food production, and the government's inability to provide basic amenities lead to the growth of urban slums. Population increases have led to increased pollution from domestic sewage, industrial development, vehicular traffic, etc. About 50 percent of people suffer from respiratory diseases in Calcutta and Mumbai, (Bhattacharya, 1997; Bhatt, 1997) and Kanpur is considered the hub of tuberculosis in India.

Population pressures also have implications for national and regional security. Environmental degradation and population increases cause the displacement of the population; migration across borders is creating potential sources of conflict within and across state lines, as well as threatening the social and political stability of the entire region (Lama, 1997). These social, economic, and political stresses are also indirect consequences of environmental stress. The population increase beyond the support thresholds brings with it a cascade of problems, which range from insecurity to civil strife.

An Overview of the State of India's Environment

This section provides a brief overview of the state of India's environment and the challenges of resource depletion and degradation that the country faces. It should be noted that given the enormity of India's landmass and the complexity of its environmental issues, we will only highlight the key concerns to give the reader a flavor of the challenges faced, rather than providing an exhaustive survey of all the problems.

Land

India is the seventh largest and the second most populous country in the world, and it has a unique physical and cultural diversity. India uses more than 50 percent of its land for agriculture. Geographical rather than population pressures account for such a high use of land for crop cultivation. Due to India's agrarian economy, land constitutes an important factor for socio-economic stability.

Due to the increasing pressures of population growth, there is a great demand for land for both agricultural and non-agricultural purposes -- for example, for industrial and infrastructure development. The overuse of land has led to the rapid degradation of what were once productive areas and are now labeled 'wastelands'. The extent of land degradation is alarming; wastelands cover about 16 percent of the country's geographical area and soil degradation affects as much as 130 to 170 million hectares of cultivable land (Parikh and Reddy, 1997).

India's land is losing its agricultural potential through soil erosion, salinization, waterlogging, and other forms of degradation, all of which cause a marked decline in soil fertility. About 27 percent of India is affected by salinization and another 27 percent by soil erosion (occurring at the rate of 75 metric tons per hectare per year). Along with erosion, Indian farmlands lose 5 million tons of nitrogen, phosphorus, potash, and major plant nutrients each year, which then has to be replaced by artificial fertilizers. Due to inadequate attention to drainage and soil conservation, desertification is occurring at a rapid rate. Floods and droughts are also recurring problems. About eight million hectares of land are submerged by floodwater each year, while drought affects 64 million hectares. Deforestation, overgrazing, faulty agricultural practices, and the constant search for firewood have resulted in landslides, flash floods, desertification and water logging. (Dwivedi, 1997).

Land degradation due to the overuse of chemicals as fertilizers, and due to poor irrigation, is a growing phenomenon. Waterlogging of land due to faulty irrigation projects leads to seepage of fertilizers and toxic chemicals, resulting in the loss of large tracts of cropland every year. This compels many small farmers to move out of their traditional habitats. In an attempt to offset this loss of land, new land is constantly converted to crop production, a practice that causes land imbalances. While the 'green revolution' has its gains, small and landless farmers often do not have the resources to invest in the technology and chemicals needed, and have therefore either been left out or even suffered at its hands.

With the increasing rural population, the per capita holding of agricultural land has fallen. By the end of the 1980s, almost three-fifths of all holdings were less than one hectare in size (Repetto, 1994). All croplands are now under stress as a result of intensive use, and constant demands for increased yield compels farmers to use more fertilizers and pesticides, adding further stress to the land. The per hectare use of fertilizers in the early 1990s in India was 60 percent higher than in the United States, whereas yield levels of most major crops were much lower. Improper and unrestricted use of chemicals in agriculture is also a health hazard (Gaan, 2000).

The privatization of state land is another problem that has tended to benefit the already wealthy. Privatization can have devastating impacts on the poorest households since in many areas of India the basic necessities are provided by the village commons -- 90 percent of firewood and 70 percent of the grazing needs of the poor are supplied by the commons, compared to the relative self-sufficiency of wealthier households (Jodha, 1986). Due to various illegal encroachments and governmental privatization of land, the area of the village commons has declined dramatically in recent decades. The programs aimed to benefit the poor have only benefited the higher strata of rural society, therefore the poor lost collectively while gaining little individually.

Pasture lands: A direct consequence of the increase in usable land has been a reduction in the availability of grazing land. Animal husbandry contributes to an important livelihood for six to eight percent of the poorest people of rural India; the impacts of this have again been severest on the poorest people. India has about one-sixth of the cattle and one-half of the buffalo population of the world. Between 1951 and 1982 the livestock population of India grew by about 43.9 percent (Ministry of the Environment and Forests, 1992). However, a decrease in grazing lands has led to intensive grazing on what remains.

It has also resulted in the migration of nomads in search of new grazing land in forest areas. While the former practice has contributed to the degradation of pastures, the movement of nomads poses problems of social tension between cattle-owners and landowners. In many cases the nomads have been forced to settle in marginal lands and engage in intensive farming activities in order to survive, thereby further marginalizing the quality of the land (Dwivedi, 1997).

Pastoral nomads are looking for alternative means to earn their livelihoods as aconsequence of disturbances in the ecological balance due to the severity of the climate, continuous deterioration of soil, dwindling natural resources, and outside invasions. Common pastures constituted a source of livelihood to a large population of Rajasthan who are affected today by social and economic uncertainties. This has forced the nomads to use available grazing resources indiscriminately (Bharara, 1999). Even there they fight a losing battle and their cultural as well as economic security seems to be at risk, as is their ancestral lifestyle.

Deserts: The Thar Desert region of northwest India extends over 196,150 square kilometers. It is a fragile region subject to the excessive stresses of frequent droughts and low rainfall. The region is quite inhospitable to the large human and livestock population. Droughts also lead to wind erosion, which forms sand dunes that often cover fertile agricultural lands and block roads and railway tracks. The most affected are the districts of Jaisalmer, Bikaner, Barmer, Jodhpur, and Jalor. However, the desert region enjoys a special status in Rajasthan because of its mineral wealth. It is the second largest producer of minerals in India. But mining in the desert has caused considerable land degradation (Mohnot, 1999). Land use planning is either poor or entirely absent here and little respect is shown for desert flora, fauna or the ecosystem. The results include increased desertification and poor biomass turnover. People are losing their homes due to the degeneration of shrubs and community-protected habitats. There is also a breakdown in food, water, and shelter systems.

A history of forced irrigation projects has added to land degradation; salinisation, in particular, turns croplands into wastelands. It is often forgotten that deserts may be barren but they are not unpopulated, indeed the Thar Desert is the world's most populated desert. This is an added pressure on an already fragile system. Recently, the nuclear explosion of Pokhran-II was carried out in the desert. While no studies are yet available of the ecological security implications of the explosion on biodiversity and habitat loss, it is fair to assume that these impacts would be considerable.

Forests: Forests are a critical environmental resource as they play an essential role in the conservation of our ecosystem by recycling oxygen, nitrogen and carbon, influencing temperature and rainfall, acting as enormous sponges to collect and distribute water, protecting the soil from water and wind erosion, maintaining biological diversity, protecting wildlife and genetic resources, and contributing to economic development. It is quite evident that India's forests are being used unsustainably. As wood supplies decline, millions of poor families who use fuelwood as their major source of energy are forced to use available substitutes such as animal dung or crop residues, which would have otherwise been returned to the soil as fertilizer. This compels farmers to use chemical fertilizers that have long-term adverse effects on the productivity of the land.

Forest lands used by local communities for extraction of fuelwood, fodder leaves, minor and supplementary food items as well as for grazing are classified under one of two official categories. 'Reserved forests' are not accessible for local use, except for what is specifically permitted by the forestry department. 'Protected forests', on the other hand, allow access for all local use except under conditions prohibited by law. As might be expected, protected forests tend to be comparatively more degraded than reserved forests (Nadkarni, 1997). For the most part the British colonists treated the forests as 'timber mines'. Unfortunately, the situation has not improved since then, and now equally extensive deforestation is taking place in the name of development, urbanization, and international trade. The high monetary stakes that accompany commercial logging have even led to the emergence of logging gangs; violence can break out between them and the forest authorities, or between the gangs and local communities. As long as forests were linked to community management, the responsibility for resource management was linked to resource use. But government controls and privatization seem to have eroded this social contract and shattered the intricate protective system that had been holding forest management together (Agarwal, 1995).

Apart from the pressures of commercial logging, the forests are also under pressure from the impacts of developmental projects, roads, transmission lines, overgrazing, transfer of land for agricultural purposes, and the demands of wood for fuel and energy. The decline in forest cover has resulted in poorer watersheds. Although the rate of deforestation seemed to have stabilized in the 1990's (in comparison to the 1970's), India still lost 548,000 hectares of forest cover in the years 1993-1995 (Forest Survey of India, 1997). This loss has had major implications for resource security in India.

The tribal communities who enjoyed the 'collective ownership' of forests for centuries now find themselves thrown out of their traditional habitats as management is taken over by state agencies. Indeed, denied the use of resources that had once sustained them, they are often reduced to 'poaching' that which was once rightfully theirs in order to survive. The springs and streams providing water for domestic use have also started drying up as a consequence of deforestation. Women have to trek longer distances in search of water, adding to the drudgery of an already difficult existence. As herders are forced to take refuge elsewhere with their flocks of sheep and herds of cattle, the new lands also come under new pressures. Social strife and conflict often lie just a few steps behind.

Biodiversity: India is very rich in wildlife. It is estimated that over 75,000 species of fauna and 45,000 species of flora can be found in India (Jairajpuri, 1991). A large amount of its wildlife treasure is endemic. India has a wide variety of fish, insects and mollusks. Among the large mammals, 79 species including the famous royal bengal tiger and the one-horned rhino have been declared endangered species. Until the 1960's, game hunting took a heavy toll on the nation's wildlife, but in spite of protective legislation, poaching and trade in wildlife products continues to this day. Depletion of India's forest cover makes the protection of its immense biodiversity difficult. India has a dense forest cover only 12 percent less than its target, 33 percent. With shrinking habitats, conservation of species is possible only if they inhabit wildlife sanctuaries. Although India has created 441 such sanctuaries covering 148,700 square kilometers, it is still not enough to contain the problem of biodiversity conservation. In order to ensure that viable populations of all kinds of wildlife are preserved as a national and international heritage, concerted efforts will have to be made by governmental and non-governmental agencies and individuals.

Water

Water is the basic building block of all life on earth. Although India has abundant water resources, these are under severe pressure and have become a scarce commodity in many areas. Over the years, water has been subjected to the various pressures of population, urbanization, agriculture, industrialization, and pollution. The over-exploitation of water on top of improper management practices has set a dangerous trend that jeopardizes the availability of water resources for future generations. Water riots have already occurred in several Indian cities and inequities in water availability and use have become sources of domestic, national, and regional insecurity.

Inter-state strategic contests over water can be attributed to a number of factors, including the degree of scarcity, the extent to which supply is shared by more than one region or state, the relative power of the basin-states, and the possibility of access to alternative fresh water sources (Gleick, 1992). An analytical appraisal of water resources *vis-à-vis* security issues is both direct and indirect. Water scarcity impacts local, state, and regional security directly, and affects the environment indirectly by triggering long-term security effects of non-renewability.

Water availability: The current water balance in India is confronted with problems of both availability and pollution. River basins are the natural hydrologic units, and the Central Water Commission has identified 24 of them in India. The per capita availability of water has reduced from 5,277 cubic meters in 1955 to 2,200 cubic meters in the year 2000. However, per capita per year water use is projected to increase to approximately 725-750 cubic meters in 2050 as compared to about 650 cubic meters in 1999 (Ministry of Water Resources, 1999).

Even though the situation seems to be in control in terms of aggregate numbers, the situation in specific areas can be dire. Different parts of the country have different problems. If the canal-fed northern areas suffer from waterlogging and human health impacts, the coastal areas suffer from influx of saline water, making safe drinking water unavailable. It is not the lack of water as much as unbalanced geographical distribution compounded with the problems of pollution that are responsible for the water crisis (Chadha, 1999).

Pollution: Toxic pollution levels in river water is a major concern throughout South Asia. Rivers provide a lifeline and serve as the primary source of drinking, agricultural, and industrial water. The major rivers of India, the Ganga and Yamuna, have nurtured civilization for centuries but have become highly polluted. Sewage and waste is often thrown directly into rivers, and even where there are municipal water-treatment facilities, they do not remove trace heavy metals and effluents from industries, often containing toxic wastes. (CSE, 1999). It is this water that is often used for drinking purposes downstream. With much water being removed for irrigation canals, the amount of remaining fresh water is not enough to dilute all the pollutants that enter the water. Consumers are also affected by agricultural contamination from pesticides and fertilizers running into the rivers during monsoons and heavy rains. Dumping of animal carcasses adds to the pollution load. Cremation at riverbanks and the religious practice of throwing the unburned bodies of sages and infants into rivers have given them an unhealthy dimension. However, the main sources of pollution are still sewage and municipal effluents, which account for 75 percent of the pollution load in rivers; the remaining 25 percent comes from industrial effluents and non-point pollution sources (Agarwal, Narain and Sen, 1999).

Groundwater demand in India has been increasing as a secondary source of drinking water, as well as for irrigation. Currently, more than 50 percent of all water requirements are met by groundwater, and more than 80 percent of all rural water comes from groundwater sources, as does 50 percent of the water used for irrigation (Chadha, 1999). The level of contamination and pollution in groundwater remains alarmingly high (Nelson, 1999) and the increased usage is triggering sinking water tables and the seepage of pollutants to deeper layers of the earth. The main causes of groundwater pollution are similar to those of freshwater contamination.

Irrigation and development projects: A number of major irrigation projects and plans have been launched for effective water management in the domestic, industrial, and agricultural sectors. However, such programs have failed to either increase availability or improve water use practices. Moreover, programs and plans in the name of development have led to massive displacements and forced migration. Major irrigation and hydroelectric projects have led to the submergence of land, mass migrations, displacement, and loss of livelihoods. In addition, traditional social relations and community networks break down, which often leads to impoverishment and insecurity. India has over four thousand 'large dams', as defined by the International Commission on Large Dams. Initially, land was not scarce and relocation was easy, so environmental and social costs were brushed aside. However, as population pressures on land increased and there arose a growing awareness of the disparity between beneficiaries and losers, mass agitations became routine -- the Narmada and Tehri projects were only the most internationally visible of those. In the context of large irrigation and development projects, the conflict between the forces of economic growth and those of human security is now recognized and well-articulated.

Water and quality of life: Eighty-one percent of the Indian population has access to safe water, but only 29 percent has access to sanitation (World Bank, 1998). Water abounds in some areas and is deficient elsewhere. While the availability of safe drinking water is an important achievement, these two facts in conjunction point towards a rather serious health problem, especially as it pertains to water-borne diseases, which account for as much as 60 percent of all urban deaths. The proportion of the Indian population that has no access to clean, hygienic, and safe drinking water is often forced to use canals and drains for both drinking and bathing purposes.

Naturally, this dual usage cannot allow maintenance of water potability standards. Ground aquifers and rivers are polluted with refuse, garbage, industrial waste, and raw sewage. India's two major Himalayan rivers, the Ganga and the Yamuna, have turned into virtual sewers. The water of the Yamuna at New Delhi is so polluted that even after prolonged boiling for twenty minutes, its foul smell perforce negates its consumption (Hinrichsen, 1998). Water supplies in rural India are contaminated with agricultural washouts containing fertilizer traces, chemicals, and pesticides. As population grows alongside the demand for water, the stress on water resources is bound to grow. Water quality issues will assume even more importance as focal points of social stress and community strife. Polluted water, water shortages, and unsanitary living conditions are all direct threats to human health and security.

Water, of course, is a critical resource. It is also a resource that holds the region together. All the countries of South Asia are tied, one way or the other, to their neighbors by water. Can we envisage a new kind of international relations built around the positive synergies of these ties? Box 2.1 attempts to initiate a discussion of integrated water management in South Asia as the basis for a new kind of regional international relations.

Box 2.1

Integrated water management in South Asia

The Himalayan water system feeds the life cycles of much of South Asia (except for Sri Lanka). Extending from the highest mountain chain on earth to the diverse ecological zones of South Asia, the rivers Ganga, Brahmaputra, and Indus are the blessings of nature and a perennial source of water for the multitude of humans dwelling in the region.

However, developmental and environmental stresses pose great challenges to this vital resource and its management (Gyawali and Dixit, 1994). This Himalayan water system is the source of one of the largest supplies of fresh water in the world and the hydro-ecology of this Himalayan system is at the crux of the discussion over water as a commodity for India and its neighbors to share (Bandopadhya, 1992).

Life and living in India, Pakistan, Bangladesh, Bhutan, and Nepal lies very much on the contours of the Himalayan water system. The Indus, Ganga, and Brahmaputra originate close to Mount Kailash and travel in different directions. The catchment areas for the Indus are shared between India (27.6 percent) and Pakistan (72.4 percent); for the Ganga, between India (83 percent), Nepal (16 percent), and Bangladesh (0.85 percent); and for the Brahmaputra, between Tibet (49 percent), India (30 percent), and Bangladesh (12 percent).

The Himalayan water system, however, works on a rhythm of its own. It has been dammed, canalled, pumped, and barraged, but each option has its limits. For example, the verticality limits spatial distribution and creates droughts and flooding in different places. Pumping can be costly, impractical, or environmentally unsound (Bandopadhya, 1992). At the same time, the growing needs of human settlements, agriculture, and industry have begun to push the limits of the system (Verghese, 1990). Efforts to tame the mighty Himalayan water system have run into the limits of our ecological, socio-economic, and technological understanding on the one hand (Murthy, 1978) and the dilemmas of hydrological uncertainties on the other (Ives and Messerli, 1989). Ill-informed intervention, guided by short-term economic goals and political interests, has created negative externalities beyond repair. Piece-meal national efforts would be insufficient, but water sharing and management should be collectively tackled by the nation-states concerned with utilization of available water in a spirit of cooperation.

However, a cooperative regional water management system is unlikely to materialize in South Asia if old ideas and lingering animosities continue to bog down the progress. China, a neighbor on the upper riparian water system, faces an acute water shortage and therefore becomes an important factor in any scheme to work out water resource development and mobilization (Josse, 1994). Could water be the issue that brings these states closer together to work on a resource that they all share and cherish? Indeed, the disagreements between the countries are certainly not trivial but they need not be insurmountable. For the most part, they are based on outmoded, neo-colonial perspectives. If an understanding between China and India allowed for water development, it could be a great boon for South Asia.

There are currently no water sharing disputes between India and Pakistan and the contentions between India and Bangladesh on the

Farrakha barrage seem to have been solved. Water sharing between India and Nepal remains contentious but there exist various agreements that could become the basis for resolving the conflicts. The challenge today is to work towards an integrated Himalayan water regime for South Asia. Creating a regional regime around such a sensitive issue cannot possibly be easy. However, water may well be the one issue on which it is possible. This is due to the cross-boundary nature of hydrologies, the regional and even global nature of the environmental crisis concerning water, and the primacy of the poverty challenge faced by all the countries of the region and which is intrinsically linked to the issue of water.

Air Pollution

The two most important global environmental concerns are those relating to global climate change and to stratospheric ozone depletion. In terms of discharging both climate change-inducing and ozone-depleting substances, India is by far the biggest emitter in South Asia. In terms of more local air pollution, the air we breathe has obvious 'security' implications for those who breathe it. The cyclic balance of regeneration enables the use and reuse of air, but nature's meticulous methods of restoring air quality have been thrown out of gear by the impinging influence of industrial emissions, automobile exhaust, toxic and gaseous wastes, densely packed habitats, population growth, and human meddling with natural systems. The factor most responsible for the deterioration of air quality in cities is the exponential increase in petrol and diesel-fuelled vehicles. Added to this are the coal-based thermal power stations and oil exploration installations. One must focus, however, not only on pollutants causing local health disorders but also on regional and global impacts since air currents do not recognize national boundaries. Cross-country air pollution and acid rain are potential sources of transboundary impacts, and therefore disagreement, if not outright conflict. Military activity, especially in a nuclear South Asia, can be yet another cause for air degradation.

Sources of air pollution in India: Air pollution mainly contains hydrocarbons and dust carbons discharged from bituminous coal combustion, petrochemical processes, and automobile exhaust. Gases like sulfur dioxide, carbon monoxide, nitrous oxide, hydrocarbon vapor, benzopyrene, and others are poisonous and toxic in nature, and extremely hazardous to humans in terms of health. A recent study from the Centre for Science and Environment (CSE, 1999) shows that for India's major cities the challenge of air pollution, particularly urban air pollution, remains daunting.

Factors which contribute to air pollution in India include domestic pollution; transportation systems; personal automobiles; power generation; industry; etc. The fact that urban centers are very often intermingled with industrial production means that more stress is placed on urban areas, creating pockets of severe air pollution that envelop communities and lead to respiratory problems, lung cancer, tuberculosis, bronchitis, and coronary diseases. Unfortunately, the old belief that developing countries do not face the problems of industrial pollution is no longer true.

Energy

Energy is important to the environment and security nexus not only because it is linked to air pollution in terms of how it is produced and how much is produced, but also because it is the critical motor of growth in modern society. The supply of oil, gas, electricity, and petroleum are areas of major concern. The energy issue has economic and environmental ramifications that sometimes even lead to military confrontations. India is one of the largest consumers of energy in the developing world. Fossil fuels account for the majority of energy output in India but their shortages are well known. There has been growth in nuclear and hydro-electrical production but it is limited. Solar energy remains partially tapped, or otherwise untapped. The demand and supply of energy, use patterns, availability, energy reserves, and a lack of energy efficiency policies are issues that not only connect with economic and industrial growth but also affect the most important social issues that touch on sustainable development.

The Indian approach to energy has been more or less based on the conventional paradigm arguing that supplying more energy into the economy can increase growth. Here, energy consumption becomes synonymous with growth and development. However, for reasons of energy security as well as environmental quality, questions can be raised regarding the advisability of embarking on a course that requires ever more energy inputs, particularly if they are to be obtained from fossil fuel imports. Moreover, the conventional paradigm of energy policy ignores options such as energy conservation, efficient use, and alternative sources. This approach has also given rise to the environment-development trap, pitting them against one another rather than placing them together in a sustainable development framework.

Sources of energy: Coal continues to be the main source of commercial energy in India, not only for direct use in industry but also for indirect energy use through power generation. Concerted efforts made in the exploration and development of hydrocarbons have led to increases in the production of oil and natural gas. However in recent years, the production of crude oil has been stagnating. Hydroelectric generation has also increased significantly. There have also been additions to India's nuclear power generation capacity. Wind power generation has also picked up during the last five years, although in real terms it remains a very small fraction of the total. Considerable emphasis has also been placed on the exploration of lignite, especially in the states of Tamilnadu, Rajasthan, Gujarat, and Pondicherry.

Among non-commercial sources of energy, fuelwood remains an important source of energy for cooking and heating in the rural and, to some extent, urban households. Out of the total rural energy consumption, 65 percent is met by fuelwood, and the total consumption for 2001 was estimated to be 162 million tons. However, the dwindling forest cover has made this an unreliable and increasingly expensive form of energy. Dung and crop residues are now increasingly used to supplement the use of wood for fuel, and animal-based energy continues to be used for agricultural operations and rural transportation.

Demand and supply of energy: Tables 2.1 and 2.2 highlight two points quite clearly. First, the demand for energy is increasing and increasing fast. Second, the trends suggest that this demand is not being met and the deficit is likely to grow. Although Table 2.1 focuses only on electricity, it is representative of a larger picture of unmet energy needs. Similarly, Table 2.2 highlights the skewed distribution of energy. Between them, the two tables point towards the potential for conflict around energy -- the future they depict is a future where the need and demand for energy grows but the availability of reliable energy does not. One can easily imagine what this might do to the price of energy and how it might impact the poorest. Scarcity, in something as basic and necessary as energy, must ultimately lead to insecurity.

Table 2.1
Electric power demand and supply (in million kilowatt-hours)

Year	Demand	Supply	Deficit	Deficit, as percentage of Demand
1992-93	305,266	279,824	25,442	8.3 %
1993-94	323,252	299,494	23,758	7.3%
1994-95	352,260	327,281	24,979	7.1%
1995-96	389,721	254,045	47,590	9.2%
1996-97	413,490	365,900	47,590	11.5%

Source: Kumarmangalam (1998)

India is not self-sufficient in petroleum products. As a result, the import dependence on oil has been increasing. The degree of India's dependency in terms of oil stood at 35 percent in 1975, and increased to 70 percent by 1984-85. In addition to petroleum imports, superior quality coal for use in the steel industry has also been imported, and these imports had touched the 10 million ton mark by 1996-97.

The distribution of primary commercial energy resources is quite skewed. Whereas the eastern region accounts for nearly 70 percent of the total coal reserves, the western region has over 70 percent of the nation's hydrocarbon reserves. Similarly, more than 70 percent of the total hydroelectric potential in the country is located in the northern and the northeastern regions. The southern region has only 6 percent of the coal reserves and 10 percent of the total hydroelectric potential, but contains most of India's lignite deposits and also has good potential for coal bed methane, oil shale, and gas hydrates.

The total energy supply (both commercial and non-commercial, excluding draught animals) increased from 82.7 MTOE (million tons of oil equivalent) in 1950-51 to 368 MTOE in 1996-97. Although the size of primary commercial energy reserves appears to be fairly large, availability in per capita terms is quite modest on account of India's large population. If no significant additions are made to reserves in the future, the per capita availability is going to further decline in the wake of the rising population.

Table 2.2
Estimated household energy demand (in tons of coal replacement)

	Urban		Rural		Total	
	1994-1995	**1999-2000**	**1994-1995**	**1999-2000**	**1994-1995**	**1999-2000**
Households (millions)	*43.00*	*51.18*	*105.79*	*109.60*	*145.79*	*160.78*
Liquefied Petroleum Gas (mtcr)	24.91	27.68	Nil.	Nil.	24.91	27.68
Soft Coke (mtcr)	7.20	7.20	2.25	2.25	9.45	9.45
Kerosene (mtcr)	15.36-31.86	30.21-46.71	3.50	3.90	16.86-35.36	24.17-50.67
Fuelwood (million tons)	6.34-24.7	24.2-57.0	139.9	159.4	146.2-164.6	183.6-216.4
Cow-dung (mtcr)	Nil.	Nil.	20.00	20.00	20.00	20.00
Crop Residue (mtcr)	Nil.	Nil.	21.50	23.59	21.50	23.59

Source: Natarajan (1990)

Unconventional sources of energy: The role of unconventional sources of energy becomes important in the backdrop of reducing per capita availability of conventional energy resources. Some interesting efforts are underway by governmental agencies and private entrepreneurs to research renewable, clean fossil fuels and cost-effective technologies. These include projects on energy from urban and industrial wastes, biomass power from gasifiers, solar power, wind energy, and small hydroelectric plants. Wind power has attained a commercial stage in many of the southern states and the increased participation of private entrepreneurs in the construction of wind farms has greatly facilitated their production. Areas where much work still remains to be done include fuel cells, photovoltaics, and the use of bio-energy. The development of energy supply options that increase efficiency and reduce polluting emissions will promote sustainable development objectives and move the country towards greater energy and environmental security in the future.

The energy policy of India has always emphasised the expansion of capacity. Subsidized prices, which fail even to cover costs, have discouraged energy-saving devices and technological innovations in existing installations. Loss and theft of energy in electricity transmissions and distribution now stands at nearly 23 percent. A major step towards energy security would be to begin focusing on energy 'efficiency' in addition to energy 'generation'. Indeed, a kilowatt-hour saved is a kilowatt-hour generated. Environmentally sound energy production and the appropriate mix of different types of energy to achieve sustainable energy consumption patterns are also important; energy prices, taxes, and other policies should be designed accordingly.

Security Implications of Degradation & Scope of Conflict

Having surveyed the state of environmental stress and degradation in India, let us now turn our attention to their security implications. Environmental degradation and its implications on human development present a new set of challenges both to conventional notions of security and to sustainable development. Mass poverty, illiteracy, disease, squalor, and malnutrition affecting a large proportion of the world's population often become the causes of social strain, tension, and strife (Fisher, 1993). The influence of this degradation can also be felt on the relationship between and among states. Conflictual relationships begin to establish themselves as simple scarcities attributed to a conflict occurring among states over some renewable resources such as river water, fish or agriculturally-productive land (Homer-Dixon, 1991). When set in an already volatile context, such as South Asia, these local and cross-border disputes can easily flare into larger conflicts.

Thus, conflicts can emanate directly from the overuse and exploitation of a resource or by the indirect effects of such degradation leading to mass migration, refugee flow, poverty, or food insecurity. This affects not only the social and economic life of the concerned state but also leads to interstate conflicts. India is no exception to this situation and, like other developing countries, is particularly vulnerable to such conflicts. In the context of the information provided in the previous section, let us explore the links between environmental degradation and security in a few selected sectors.

Land and Security

Land has been, and still remains, a major cause for strife and conflict at every level. In India, land is under stress for a number of reasons, including the dwindling availability of arable land, adverse impacts of intensive and capital-intensive agriculture, loss of soil fertility, problems related to irrigation, the fragmentation of holdings, rural migration, etc. These processes result in synergistic impacts on agricultural productivity, marked declines in crop yields, and a per capita reduction of land, creating social disparity and leading to civil strife and human insecurity. Government control over forests definitely resulted in the reallocation of forest resources away from local communities, transferring it into the hands of urban and industrial elites. This increased social conflict and the destruction and transformation of the ecological resource. Similarly, increasing cattle populations, overgrazing, and ecological changes have made it difficult for nomadic pastoralists to continue their lifestyle. This lifestyle was, in fact, well suited to the arid and semi-arid conditions of their surroundings but is now under threat from the pressures of development and modernity. With cash cropping and an increasing use of chemical fertilizers, the synergistic ecological relationship between the nomadic grazers and settled farmers has changed into one of conflict. More families now look for sedentary lifestyles. The worst clashes and conflicts to date have been over land, and many outstanding territorial claims still persist.

At its root, the human-to-land ratio continues to decline as populations rise, while the land availability remains more or less static. The growing scarcity of land has serious implications for the social, economic, and political relations among various communities. Worsening socio-economic disparities within the state and an inability to meet the land demands of a growing population have the potential for immense conflict.

Water and Security

As we noted in an earlier section, there are multiple causes of water insecurity including drought, the spread of deserts, change of river courses, uneven rains, acid rain, non-availability of hygienic potable water, deforestation, climate change, and other reasons. At its worst, and this is not uncommon, water insecurity can either lead to severe health problems or force people to migrate in search of water (which puts added pressure on the resource). However, the uncertainty of climate change perforce leaves no place that is permanently safe. The recent drought in Gujarat or the wild growth of the sea at Vishakhapatnam have been indiscriminate in their impacts. Indeed, the evidence suggests that the South Asian water system, both coastal and freshwater, is likely to be severely affected by global climate change.

Indeed, the multiple stresses on water, both in terms of quantity and quality, have already caused individual conflicts, group clashes, civil strife, interstate conflicts, and regional confrontations. At the micro level, water-deficient community pumps facing long lines of pots needing to be filled is a common sight that spells both verbal and physical abuse in most of the cities and towns of India. The major rivers of the country are almost all 'multinational' rivers. Increasing demands lead to interstate disputes over the sharing of water. Some of the most obvious examples of such disputes include the Godavari water dispute among the states of Maharastra, Andhra Pradesh, Karnataka, Madhya Pradesh, and Orissa; the Krishna Water dispute, involving the states of Maharastra, Karnataka, and Andhra Pradesh; the Narmada basin water dispute involving the states of Maharastra, Gujrat, and Madhya Pradesh; the Cauvery water dispute between the states of Karnataka and Tamil Nadu; and the Ravi-Beas water dispute between the states of Punjab, Haryana, and Rajastan. Conflicts over fish productivity, oil, gas, oceanic wealth sharing, migrating routes of ships, strategic contents of seas, and the non-demarcation of maritime boundaries are other areas causing tension and conflicts, putting human and state security at risk.

Water-related conflicts can also arise from the secondary impacts of developmental schemes including irrigation, hydroelectric production, and flood control reservoirs. Major water development projects can lead to environmental degradation, the submergence of land, the environmentally-induced migration of refugees, and rehabilitation problems like those of Silent Valley, Bodhghat, and Bethi projects, as well as the Tehri Dam and Sardar Sarovar projects. As other chapters in this volume testify, agitation and even conflict over such projects is now a fairly routine event not only in India, but all over South Asia.

Transboundary rivers flow from, into or through six of India's neighbors, including Nepal, Bhutan, Tibet/China, Myanmar, Bangladesh and Pakistan. This geopolitical reality can neither be ignored nor reversed. Flood-induced migration from Bangladesh has caused the worst-ever clashes between Bengali migrants from Bangladesh and the local populations in Assam and Tripura, due to pressure on local resources and the changed character of the states' ethnic and cultural profiles. When rivers form international boundaries, as in the case of India and Bangladesh, flood protection measures on one side can result in inundation on the other side, spurring a loss of land and resources. Problems also arise when rivers flow from one country to another, as was the case in the Ravi-Beas rivers between India and Pakistan. The conflicts in these cases are clearly international security issues and will remain an essential part of the region's geopolitics, given the geography of South Asia and its waterways.

Other conflicts have occurred over fishing rights between local fishermen and the mechanized trawlers of corporate companies in the coastal areas of India, and problems of over-fishing between India, Sri Lanka and Bangladesh. These conflicts pose imminent threats to the food security, economic security, and ecological security of the regions and countries in question, and can also take on international security dimensions. Disputes related to fisheries might well become more frequent as nations harvest fish stocks beyond the level of sustainable yields.

Potential sources of conflict also exist in the threat of global climate change, which is likely to impact water availability in countries such as India. Rapid rises in evaporation can affect soil fertility, flood plains, and mountain ecology, and the risk of floods, landslides and storms may increase. This would have strategic implications for India and its neighbors.

Energy and Security

The energy demands of India are increasing with its population and prosperity. The centrality of energy resources to life and livelihoods is such that competition and stress on this critical resource could potentially lead to conflicts, protests, tensions, unrest, riots, and even armed confrontation and wars. Each one of these has, in fact, happened in the very recent past in India, South Asia or elsewhere. Energy is therefore among the most volatile and potentially dangerous links in the environment and security complex.

Micro-conflicts over fuelwood are already common at the local level, but confrontation, or at least competition, over fossil fuels exists even today. Insecurity and tension over energy issues already exists between India and its neighbors and is manifest in the politics of hydropower generation on shared rivers as well as in heated discussions about transboundary gas and petroleum pipelines. These debates are very much in the realm of the contemporary security discourse and are likely to take even more prominence in the future.

At more local levels, vehicular pollution in cities such as Bombay, Calcutta, Bangalore, and Chennai is causing respiratory problems, eye inflammation, skin allergies, etc. and is directly endangering human security. At the same time, regular cuts in the amount of available urban electricity can literally paralyze individual life and the national economy. Indeed, the ramifications of economic insecurity caused by the energy crisis can be very far-reaching and could include indirect but painful impacts, including a lack of productivity, unemployment, and the attendant issues of labor protests, etc.

The energy challenge cannot be solved simply by infrastructural solutions focused only on supply. Energy security is not only a function of providing the energy needed, it is also related to the national and individual costs of accessing this energy as well as the related health and environmental impacts. Dependence on external sources for this critical resource has sovereignty and security implications of its own. Policies aimed at energy security should therefore focus on the amount of energy available as well as the energy mix, source, and impacts.

Conflict as a Cause of Environmental Insecurity

Environmental insecurity is both a cause and an effect of political tension and military conflict. These conflicts involve high costs and need the diversion of scarce resources in order to sustain the spiraling arms race. What is less appreciated is that regional cold wars cause the destruction of natural resources and degradation of environment on a gigantic scale. For example, the nuclear tests of May 1998 were a wake-up call signaling that twenty-five years of peace in the region may well have ended. Apart from the actual fear of a regional nuclear holocaust and its effect on human life, nuclear weapons create serious environmental ramifications. They generate toxic waste that is difficult to dispose of. Storage sites pollute the environment of the region by adulterating land and water. Large tracts of land are requisitioned and put to non-productive uses, such as the construction of test sites, firing ranges, depots, waste dumps etc. Long-term ecological implications of militarization and nuclearization have deep impacts on human health and security.

Arms competition and conflict also create major obstacles for sustainable development. They make use of the scarce material resources required to support conflicts. They pre-empt human resources and wealth that could be used to combat the collapse of environmental support systems, poverty, and underdevelopment. Conflict and confrontations stimulate traditions of non-cooperation between nations whose ecological and economic interdependence requires overcoming national or ideological antipathies. Conflict results in loss and damage of land. Up to one percent of all the land in the world is directly used by militaries (ICHME, 1991). Chemical and biological warfare pose enormous environmental hazards to the atmosphere. The existence of nuclear weapons and the destructive potential inherent in the intensity of modern warfare have given rise to a new dimension of security. The pollution due to firepower and degradation of the natural environment in the Kargil-Drass sector in Kashmir points in this direction. Similarly, damage to the natural environment at Siachen, which lies in the higher reaches of the Karakoram mountains, occurs due to inhabitation by the military defending the border between India and Pakistan (Singh, 1999). In the nuclear age, nations can no longer obtain security at each other's expense; they must seek it through cooperation, agreements, and mutual restraints instead of deterrence and the threat of war.

Interdependence is fundamental in terms of the environment and economics, and in the contemporary world it is also beginning to encompass military security. However, environmental stress increases the prospects for low-intensity conflicts in South Asia. These conflicts are likely to escalate due to scarcity of resources and increasing competition to acquire them. The situation in the region presents particular dangers in the context of environmental militarization and poverty-induced stresses. Large-scale movements of refugees, competition for scarce water and fertile land, deposits of oil, and ill-defined boundaries add to the existing tensions and increase the possibilities of conflict. South Asia's import of arms places enormous demands on energy, mineral resources, and capital. The misdirection of scarce resources to an arms race contributes to further insecurities for society in general and the nation at large.

Implications for Regional Security

Environmental issues will continue to demand greater and greater attention in South Asia. The challenge that faces the region today is reconciling the huge population with the finite natural resources of the region. The worst clashes to date have been fought over land, and conventional clashes over simmering territorial claims on Kashmir and Siachen have triggered actual wars between India and Pakistan.

Although water has not been the cause of inter-state conflict in the military sense, it has been a major source of regional discord. The nations of this region are under enormous pressure. In India's case the unquenchable thirst for more water arises from the issues of population, industrialization, urbanization, and large-scale agriculture. Pakistan, with 78 percent of its land under irrigation, fares no better. In Bangladesh the problem of floods continues to grow worse. Water-sharing agreements have been, and will remain, a major challenge for the nations of the region. However, these agreements can be viewed as potential opportunities for cooperation instead of simply as threats.

Environmental issues become identified as threats to international or regional security when they undermine the social, economic, and ecological health and well-being of neighboring countries. For example, industrial pollution from the south is destroying the fragile mountain ecology of Nepal, thereby undermining its tourist industry, a major source of foreign exchange earnings. Unilateral withdrawal of water from international rivers causes havoc to agriculture in a downstream riparian state. Radiation leaks from nuclear testing are carried by wind into neighboring states, thereby affecting the health of their nationals. Regional security is compromised not only because of actual environmental impacts but also because of the toll such discord and conflict takes on mutual cooperation and goodwill in the region.

Environmental stress is seldom the only cause of major conflicts within or among nations. Nevertheless, these conflicts can arise, often in conjunction with other issues, from the marginalization of sectors of the population and the ensuing violence. This occurs when political processes are unable to handle the effects of environmental stress. Environmental stress can therefore be an important part of the web of causality associated with any conflict and can in some cases be catalytic. A manifestation of the growing concern in the international community is the phenomena of environmental refugees. The immediate cause of any mass movement of refugees may appear to be political upheaval and military violence, but the underlying causes often include the deterioration of natural resources and their capacity to support the population. As unsustainable forms of development push individual South Asian countries up against environmental limits, major differences in environmental endowments among countries can exacerbate international tension. Competition for the use of the global or regional commons such as the oceans and fisheries, or for the use of more localized common resources such as rivers and coastal waters can escalate to the level of regional conflicts and threaten regional peace and security.

Environmental stress can create conditions where political processes are unable to handle its effects, resulting in political upheavals and military violence. Poverty, injustice, environmental degradation, and conflict interact in complex and potent ways. The problems of South Asia present a grim scenario of environmental resource exploitation because of accompanying increases in population, poverty, unemployment, social chaos, and political instability. While there is a great deal of effort in the areas of growth and development, the adverse economic trends, misplaced development policies, and inequitious multi-racial and multi-ethnic societies complicate the linkage between the environment, development, and security, making the region even more conflict-prone and less secure.

Regional Cooperation for Peace & Security

The first step in creating a shift in the region's perception of its security challenges and opportunities is to broaden our vision. Conflicts may arise not only because of political and military threats to national sovereignty, but also be derived from environmental degradation and the pre-emption of development options. There are, of course, no military solutions to 'environmental insecurity'. The very idea of national sovereignty seems to be undergoing fundamental modification by the pressures of economic, environmental, and cultural interdependence. The global commons cannot be managed from any national center, and nation-states are ill-equipped to deal with threats to their shared ecosystems. The challenge for South Asia in terms of environment and security is really a challenge of cooperation. To the extent that sustainable development provides the joint focus on environmental and development issues, it also provides the template around which South Asia can begin building cooperative arrangements to deal with these twin issues of joint concern and priority.

South Asia shares common rivers, common watersheds, and common mountain ranges. Consequently, deforestation in one country can cause flash floods and soil erosion in another. Dumping untreated industrial waste into a river upstream can cause consequent toxic effects on fisheries and forests downstream in another country. Major wind currents move west to east during winter, across the national boundary between Pakistan and India, while in Sumatra the monsoon winds move in the opposite direction from India to Pakistan (Hussain, 1990). A thermal plant releasing untreated exhaust in one country can, therefore, render the air unhealthy in the other. Such transnational interconnections among the countries of the region only emphasize the need for regional cooperation in South Asia. Areas of such cooperation might include joint efforts at the reforestation of shared watersheds, toxicity management in shared waterways, information exchange for better riparian management including flood forecasting and control, joint development of the Himalayan water system, information exchanges on local practices which may be transferable in the region, or joint research and enquiry into critical areas such as earthquake forecasting and management.

On the human development front, despite South Asia's impressive GNP growth rates over the last 50 years, the nations of the region still have very large populations of people living in abject poverty. Human mortality rates and the high frequency of diseases under deprivation of basic necessities call for working out joint strategies among South Asian nations. Unhygienic drinking water and inadequate diet result in the stunted physical growth of a large number of South Asian children. Human resource development and quality of life will be denied unless access to education is ensured and concrete steps toward human development are taken. These are problems that each South Asian nation faces individually, but they are also problems that they all share collectively. Moreover, the problem of chronic poverty is so critical to each nation in the region that working together in addressing it makes much sense.

Another area that is ripe for regional cooperation is that of an early warning system for ecological disasters. Like ecology itself, ecological disasters do not respect national boundaries. This is another key area where actual progress can be made by initiating joint regional programs of cooperation. It is of the utmost importance that governments and people are aware of imminent environmental stresses *before* they cause irreversible damage to core national interests. It would be just and fair if appropriate international and regional organizations pooled their resources to establish a reliable early warning system which would take remedial measures to control imminent environmental risks and stresses. Such systems would monitor indicators of risks and potential disputes, for example, in areas of soil erosion, flood management, growth of regional migration, and commons overuse as a threat to sustainability. The organizations may also offer services helping countries to establish the principles and institutions of joint management.

Sustainable Development as a New Environmental Security Paradigm

National security and self-preservation are the first requisites of an independent nation. Traditionally, India's security has revolved around two important factors: first, the impact of external factors on the region and the response of India and Pakistan, its two major actors; and second, insecurity resulting from the region's domestic dynamics. Due to historical, cultural, and ethnic reasons, turbulence in one part of South Asia tends to spill over into the rest of the region. This underscores the need for a regional focus on security issues.

Equally important is the need for a more inclusive definition of security. As an emerging concept, environmental security has strong links with security at multiple levels, including individual human security, national security, and regional security. A logical and sequential chain of causality is evident in the transition from human activity to environmental change, from environmental change to social disruption, and from social disruption to conflict. Many of the most pressing challenges of today, and even more of the potential threats of tomorrow, are linked to environmental concerns. It is fair to assume that conventional security debates will have to incorporate and respond to these new and emerging concerns. Sustainable development provides one possible framework for analysis of security questions within the context of the twin imperatives of economic development and environmental quality.

Efforts to generate rapid growth have damaged the natural systems and affected not only the present generation but also prospects for future generations. Intensive cultivation, clearing forests, filling wetlands, industrialization, and urbanization propelled partially by population growth eventually impair the ability of renewable resources to regenerate themselves. Development experts recognize the destruction of natural resources for short-term gains as potentially disastrous for future generations. The alternative lies between meeting the human needs of today while preserving the natural productive capacity of tomorrow, or sustainable development. This establishes the interconnection between the environmental and economic dimensions of security by redefining developments in terms of their compatibility with ecology for present and future needs; that is, redefining in terms of sustainable development (WCED, 1987). Indeed, there are practical problems in implementing sustainable development, including the political and technical challenges of measuring sustainability, changes in economic policies, and the interests of powerful elites, among others.

However, the sustainable development approach is so deeply rooted in the concept of human security that it becomes a powerful metric within which to analyze issues lying at the nexus of environmental and security-related concerns.

What would policies within a human security-focused, sustainable development framework look like? This, of course, is a question for further research and analysis. However, some of the answer should be evident from the discussion above. It would, for example, shift the focus from large infrastructural solutions to local and water resource development and management. It would take an inter-basin approach to regional water management. It would place as much emphasis on environmental impact assessment as is now placed on assessments of economic costs and benefits. It will place emphasis on population growth in the contexts of human security and environmental quality. And finally, both policy making and policy implementation would be based on institutions and governmental structures that are participatory and democratic, emphasizing relevant stakeholder and community knowledge rather than simply following top-down edicts.

The compulsions of environmental paradigms now have deeper and larger dimensions in the development of the regional economy, industrial growth, trade, and national power. The notion of 'environment' is so broad as to include all aspects of policy as well as all aspects of security. However, the environment in South Asia is vitiated by mutual mistrust and a psyche of suspicion that clings to problems that are ultimately of little consequence to the prosperity of the region and its people. The need for a fresh look at the issue of exactly what we mean by 'security' is, therefore, very much needed.

Admittedly, the environmental stress is directly proportional to increases in the prospects for low-intensity conflict within South Asia. Nations have often fought to assert or resist control over raw materials, land, river basins, waterways, energy supplies, and other key resources. These conflicts are likely to escalate in the face of resource scarcity. Nations like India who have massive resources, both human and environmental, may survive the compulsions of environmental onslaughts, although for nations at the brink of economic and environmental collapse, the scenario is much murkier and grimmer.

However, all nations of the region will benefit from a diversion of resources to human development, poverty reduction, and environmental enhancement. Although there is a lesser probability of direct war over environmental resources in South Asia, the scarcity of resources would create situations of domestic turmoil, which could easily spill across to other nations in the region. The urgency of seeking joint solutions to common problems becomes all the more pressing as South Asia continues to grapple with the challenges of globalization and of economic, social, and environmental integration.

The traditional approach to security, with its primary focus on military power, must be broadened to encompass the real sources of insecurity – unsustainable development as an aftermath of environmental exploitation, resulting in a deepening of the traditional forms of conflict. The underpinning dictum of the time must be cooperation, coexistence, equality, and mutual benefit as enshrined in the doctrine of 'Panch-shila' – the five key principles of Indian foreign policy (mutual respect for territorial integrity and sovereignty, mutual non-aggression, mutual noninterference in internal affairs, equality and mutual benefit, and peaceful coexistence) that Prime Minister Jawaharlal Nehru enunciated in 1954.

Chapter Three

Environmental Security in Pakistan: Are there Grounds for Optimism?

SHAHEEN RAFI KHAN

"The situation is epitomized by the leader who proclaims he will not permit one square meter of national territory to be ceded to a foreign invader, while allowing hundreds of square miles of topsoil to be eroded away each year."

-- **Unknown**

ENVIRONMENTAL SECURITY IS a still-evolving notion that covers a vast area with diffuse causes and heterogeneous impacts. In order to study environmental security -- or rather insecurity -- in the context of Pakistan, this chapter will first briefly review the competing perspectives on environmental security. The next two sections will root the discussion in Pakistan's development experience, initially outlining the paths to environmental degradation and then assigning the impacts of it on the human security of Pakistan's poor. This will be done in the context of sustainable human development criteria, focusing in particular on the resultant environmental impacts, the growing vulnerability of the poor, their responses, the induced insecurity, and conflict potential. A simple environment-security model, which is the formal setting for this model, will also be explained. Finally, a case study will be presented to highlight the application of the concepts developed here with a focus on the institutional prerequisites for transparent and effective governance.

Perspectives on Environmental Security

The Conventional View

Conventionally defined, the term 'security' usually refers to the defense of sovereign states against violent attack, either from other states or from terrorist or revolutionary groups within its borders. It evokes an image of soldiers, missiles, tanks, guns, bombs, with states, either individually or in alliance, as equals, or in a patron-client relationship, perpetually ready for military confrontation. Clearly, this impersonal and organized aspect of security leaves little space for the concerns of ordinary people, who also seek security in their ordinary lives.

The tendency to fold emerging environmental concerns into this narrow military context is, seemingly, not misplaced. Competition for land and natural resources (e.g., oil, minerals, etc.) has long been a source of hostilities between nations. These resources are seen in strictly strategic terms as enhancing a state's ability to protect itself from aggressors and competitors. Equally, oil, coal, and fuel factories are vital for tanks and naval ships, contributing to states' capacities to wage war and achieve security from violence. But this is where the parallel ends. Non-renewable resources lend themselves more easily to conflictual mindsets, whereas renewable resource-related problems tend to transcend state boundaries and require cooperative solutions. Furthermore, resource wars are becoming a diminishing possibility. Globalization and the ascendancy of global trade clearly prefers market, rather than military, solutions. Small nations are protected by international law and are increasingly able to defend themselves against armed incursions. Finally, thanks to technology, cheap substitutes are available for virtually every raw material, except oil.

The attempt to seek a convergence of perceptions should, perhaps, defer to the functional dissimilarities. Thus, war destroys the environment directly, as exemplified by the wartime nuclear detonations by the US in Japan, or its use of defoliants in Indochina. In fact, preparation for war, for instance nuclear testing and biological and chemical weapons research, is equally, if not more, destructive of the environment, consuming vast fiscal, organizational, and leadership resources, which could better be spent on environmental restoration.

Human Security

Defined in more generic terms, security connotes conditions that make people feel secure against want, deprivation, and violence. People desire protection, not only from external and internal violence but also from the threats of disease, hunger, unemployment, crime, social conflict, political repression, and environmental hazards. According to the United Nations definition, "ultimately, it is a question of human security -- a child who did not die, a disease that did not spread, a job that was not cut, a dissident who was not silenced. Defined in this manner, security represents not only protection from violence and deprivation but the enhancement of life's choices, of improvements in the quality of life." (UNDP, 1994).

There is a preventive, integrative, and inter-generational aspect to human security, which contrasts sharply with its insular, aggressive and myopic national counterpart. Human well-being is embodied in a development process that is efficient, equitable, and sustainable. Its absence creates social excesses, economic problems, and environmental hazards. These also have a tendency to spill over national borders, taking diverse forms, such as famine, disease, pollution, drug trafficking, terrorism, etc. Thus, sustainable human development is not only a *sine qua non* for the security of individual nations, but it also has a cross-border aspect, where the rich nations have an obligation to help the poorer nations. In doing so they also guarantee their own security.

Environmental Security as a Subset of Human Security

In the first place, the overall ecological concern is with anthropogenic, as opposed to purely naturally-sourced, or biocentric, environmental changes. "The essence of this global change problematique is that human beings, by virtue of their numbers and the magnitude of their activities, are causing bio-geochemical changes in the Earth system that are taking place many times more rapidly than those … occurring naturally" (Soroos, 1991).

The term 'activities' can be viewed as a catch phrase for unsustainable development processes. Whether they take the form of excess consumption, as in the north, or inequitable and resource extractive development outcomes, as in the south, both contribute to environmental degradation and pollution. Also, both have global ramifications. In the first case, these are directly evident, for instance, in ozone layer depletion and global warming. In the second case, they manifest themselves through induced human behavior, such as mass migrations and the spread of contagious diseases across countries. An additional aspect of the latter is the vicious, downward spiral of poverty and environmental degradation. As Lester Brown, president of the WorldWatch Institute puts it, "Unless we redefine security, recognizing that the principal threats to our future come less from the relationship of nation to nation and more from the deteriorating relationships from ourselves and the natural systems and resources on which we depend, then the human prospect could be a bleak one. If we do not act quickly, there is a risk that environmental degradation and social disintegration could begin to feed on each other." (Myers, 1993).

This condition also draws attention to the fact that the poor are perpetrators only in their capacity as victims. Stated differently, degradation is rooted in unsustainable development. Poverty is both a direct and related outcome, in the sense that communities are further impoverished and disempowered by the degradation such development engenders. It follows, then, that poverty-induced degradation can be seen as an inescapable response, rather than as a deliberate or voluntary act (Khan and Naqvi, 2000).

Responses to Environmental Insecurity

Despite the fact that the rationale for attributing traditional security mindsets to non-renewable resources is diminishing, there is still a tendency to invoke them, even in the case of renewable resources. For instance, protagonists argue that cross-border tensions over deforestation, pollution, water, and marine fisheries can snowball into full-scale military conflicts (Diehl and Glenditsch, 2001). Careful scrutiny reveals that aggressive posturing notwithstanding, bilateral (such as the Indus Basin Treaty between India and Pakistan) and multilateral accords (such as those on marine fisheries) are more the norm. The prospect of water accords between Jordan and Israel, Iran and Iraq, India and Bangladesh, and Nepal and India appear likely at some point in the not too distant future.

However, agreements risk unfolding unless they are underpinned by appropriate domestic policies for efficient and sustainable resource use. In fact, the risk of military confrontations increases in direct proportion to the profligate use of renewable resources. "Ultimately, the key objective should be to conserve the resource in order to maintain adequate supplies well into the future, rather than trying to control more of a resource that is being depleted" (Porter, 1995). Where the threat is transnational (ozone depletion, climate change, biodiversity loss, etc.) the logic of global cooperation is demonstrated even more vividly. The problem originates collectively and has global ramifications. Defined thus, it lies beyond the scope of established diplomacy and international relations. As Ullman (1983) puts it, "We can not launch fighter planes to resist global warming."

The advantages of responding to these problems through cooperation, rather than confrontation, are apparent. For one thing, it converts a zero or even negative-sum game into a win-win situation, where all countries can share the benefits of resource conservation and reduced pollution. Second, global responses represent the antithesis of stark Hobbesian choices where, for instance, citizens may have to opt between personal liberty and freedom from gratuitous or organized violence. In the wider context of human security, all aspects of liberty become equally important. A cooperative approach makes it possible to accommodate all these aspects, as opposed to forcing trade-offs.

An encouraging, post-Cold War development is that successive US administrations have become aware, at least in their rhetoric, of the need to redefine national security in order to encompass environmental threats. Under the first Bush administration, a 1991 presidential document defined US national security objectives to include assuring the sustainability and environmental security of the planet. The Clinton administration integrated environmental security even further into its national policy agenda. President Clinton described a stark vision of a future world filled with overpopulated countries, depleted resources, and extreme divisions of wealth and poverty, and called for a strategy of sustainable development as a comprehensive approach to the world's future (Porter, 1995). A logical corollary of this is that it is better to empower poor countries economically, politically, and socially in the interests of preserving democratic value systems and world trade rather than viewing them as client states whose loyalty can be bought by arming them to the teeth. Unfortunately, this rhetoric risks coming unstuck, given the present US administration's belligerent attitude on a myriad of issues ranging from climate change to nuclear strategy.

Defined in this manner, environmental security has certain organizational implications as well. Security as a human value is biased toward preservation of the status quo, in the sense of minimizing human-induced changes to the environment that would degrade and disrupt it, causing adverse consequences for current and future generations. In other words, "it is a conservative value orientation in regard to maintaining earth systems" (Deudney, 1990). In contrast, environmental stewardship requires behavioral and organizational changes. Its professional ethos is 'husbandmanship', or a more respectful cultivation and protection of plants, animals, and the land. It calls for partnerships between governments, NGOs, and the private sector as well as between nations and in regions.

Globalization and the North-South Debate

The downside of such a global approach is that it can easily transmute into pernicious arrangements that generate inequitable and environmentally harmful outcomes for developing countries. Thus, trade and patent agreements have engendered unsustainable resource use, land degradation, and biopiracy. The recent agitation in Seattle at the World Trade Organization (WTO) Ministerial meeting is testimony to the fact that even northern progressive elements found the WTO conditions unacceptable. Global environmental accords such as the Kyoto Protocol can have equally pernicious implications. In its present form, the Protocol could limit low-cost development options for third world countries. Although polluting, these options are deemed acceptable given the relatively low per capita emissions of these countries. Furthermore, developing countries are also encouraged to compete for funding for low-cost mitigation options through mechanisms such as the Clean Development Mechanism (CDM) envisaged by the Kyoto Protocol, thus limiting their choices at a point in time when their greenhouse gas (GHG) emissions actually converge to the global average (Agarwal, et al., 1999).

An economic interpretation offers hope for justice in the future. Ultimately, the choice lies between what individual countries (or groups of countries) have to forego in the short term in return for what they stand to gain in the future through collective action. At first glance, this may appear to be a simple economic trade-off, and somewhat prosaically, the choice is a function of the discount rate applied to collective action. Illustratively, the proclaimed success of the Montreal Protocol on ozone-depleting substances and the relative failure of the Kyoto Protocol on climate change reflect the varying costs of abatement and mitigation. As the costs of not implementing these options rise (for example, global warming and rising sea level), the rationale for collective action will grow. However, economic considerations alone cannot convert perverse incentives that promote degradation into a set of environmentally-appropriate ones. These need to be supplemented with political will, awareness and advocacy.

A Digression: Can the Military/Security Agencies be the Agents of Conservation?

A recent study on environment and insecurity in the Northwest Frontier Province (NWFP) of Pakistan states that "major military and security agencies...have a reasonably defined role that may expand their responsibilities beyond those of traditional security concerns" (Hanson,

1999). The preamble alludes to the risks incurred by viewing environmental concerns through a traditional security prism. In South Asia the incompatibility becomes more pronounced for three reasons.

First, security forces impose the will of the state unquestioningly. Sensitizing them to environmental security concerns can prove to be a double-edged weapon. Experience has shown that a liberal dispensation at the concept stage tends to metamorphose into an arrangement for imposing the will of government agencies and vested interests upon local communities. A classic example of this is when security forces resort to punitive measures when communities encroach on forests to meet their subsistence needs but overlook major acts of degradation perpetrated by timber contractors and government development agencies. Second, budget appropriations for the military compete directly with allocations for social services and environmental conservation. Essentially, the message is that fiscal reallocations, which address the problem at its source, are far more effective than enforcement, which deals with the consequences of not doing so. Third, cross-border adventurism contributes directly to environmental degradation and also through population displacement. The country paper from India in this volume dwells on this at some length. On the whole, it would not be far from the truth to state that the military in developing countries tends to be more part of the problem than part of the solution. Any discussion of their involvement in an expanded security paradigm that includes environmental security concerns needs to begin at least with a cautionary stance.

A Simple Model

In order to accommodate these ideas into a conceptual basis for further analysis, a simple environment and security model for Pakistan is presented in Figure 3.1. It is adapted from the NWFP-specific model developed by Hanson (1999), and combines features from Homer-Dixon (Deudney and Matthew, 1999) and Gleditsch (1998).

Subsequent sections will highlight two situational and case analyses to illustrate the route to conflict. The basic message is that while it takes time for insecurity to degenerate into conflict, and environmental resilience and social tolerance tend to be high and can withstand considerable stress, the long-term risks of conflict are very real in the absence of political and institutional reforms.

Figure 3.1
Insecurity and sustainable human development: A simple model

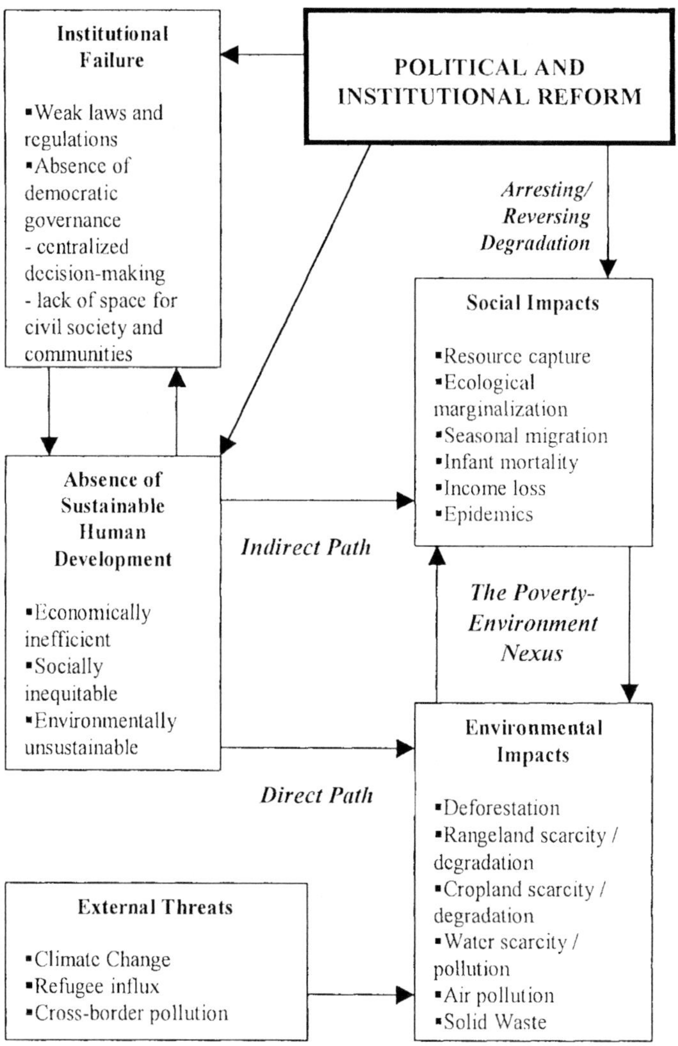

Simply put, the pathways to degradation and conflict are as follows. The absence of sustainable human development leads to environmental degradation and social inequity and injustice. The latter condition increases the vulnerability of the poor to degradation. In turn, the poor are driven to prey upon the environment. This process sets into motion a vicious downward spiral referred to as the 'poverty-environment nexus'. External influences can also be environmentally destabilizing.

The case studies discussed later in this chapter show how inadequate remediation mechanisms can result in conflict. It is almost tautological that political and institutional reform are a *sine qua non* for more environmentally and socially benign development, for arresting degradation directly and for putting in place mechanisms to prevent and mediate conflict.

Pakistan: Paths to Environmental Degradation

The previous section has defined environmental security as one of the many inter-related factors that contribute to human security. These also include economic, food, health, personal, community, and political security. The fulfillment of these conditions assures, at a minimum, a level of personal, social, and political well-being which allows people to both exercise and improve their life choices. Sustainable human development, or SHD, is both a necessary and sufficient condition for this, in as much as it establishes the conditions for human security as well as the means for sustaining and enhancing it.

Briefly, SHD is an embracing concept which defines and integrates development, equity, and sustainability in relation to the three central pillars of sustainable development -- economic, social, and environmental (Munasinghe, 2000; Khan, 1998). Table 3.1 illustrates the various ways in which Pakistan has fallen short of meeting minimum SHD criteria. The remainder of this section will focus on those aspects of the development process that have engendered particularly adverse environmental impacts. The next major section will investigate the impacts of this degradation on the poor.

Table 3.1
The absence of sustainable human development (SHD)

	Economic	Social	Environmental
Development/ Efficiency	Macroeconomic instability (fiscal deficits, current account imbalances, exchange rate overvaluation) Sector price distortions (industrial protection via tariffs and licenses, input and output price subsidies, degrees of state intervention in industrial production, utilities service delivery, internal and external trade)	Low social sector allocations (education, health, water supply and sanitation, income generation, employment) Poor outreach of poverty reduction schemes, politically influenced entitlements and misappropriation of funds	Policy neglect, minimal allocations for natural resource conservation and the use of environment-friendly technologies Failure to internalize environmental effects (perverse incentives encouraging extraction rather than conservation, inability to enforce/comply with environmental emission and pollution standards)
Equity	Tax policies favoring the affluent (regressive indirect taxes, agricultural income tax exemptions)	Minimum provision for grassroots participation and empowerment of disadvantaged groups in the formulation and implementation of policies and programs Centralized decision-making provides little space for social capital to develop or articulate itself	Degradation and pollution impact disadvantaged groups adversely (ecological marginalization, industrial and vehicular pollution, household fuel consumption, sanitation, and waste disposal)

Sustainability	Degradation of physical infrastructure (industrial, municipal, transport, and utilities)	Lack of resilience in political systems (military coups replacing elected governments) Withering of informal legal and social systems including support networks and resource management practices due to modernization	Degradation -- irreversible in many cases -- and increased vulnerability (deforestation, land degradation, biodiversity loss, coastal zone erosion, air, soil, and water quality deterioration)

The Institutional Vacuum: A Direct Route

The existing state of degradation and pollution is a result of the failure to mainstream environmental activities within the development process (Jalal, 1993). Specifically, this refers not to the management of the environment *per se*, but to the management of development activities within the assimilative capacity of the environment. At the risk of some generalization, one might define this process as highly resource-, capital-, and technology-intensive. Relatively little weight is given to economic and financial incentives as a means of conserving land, forest, air, and water resources, or using them in a sustainable manner.

A more critical problem lies in how this development is managed. The prevailing trend in Pakistan has been for a nexus of landed, industrial, military, and techno-bureaucratic elements to give its own shape and form to the prevailing development paradigm, and to capture most of its benefits. Modernization in and of itself contains impulses towards degradation. For instance, when economic opportunity and mobility depreciates the value of the resource base for communities drawing their sustenance from it, or when national legal and regulatory systems override traditional, community-based resource management, resources are exploited unsustainably.

When development takes place in the absence of democratic and decentralized governance, environmental problems are compounded. Centralized state institutions tend, at best, to be insensitive to the importance of community property rights and empowerment and, in general, to the need for public space to facilitate the growth of social capital. At worst, they are hostile to decentralized initiatives, viewing them as a threat to their sovereignty. This is unfortunate as, backed by the right incentives, traditional community practices offer considerable hope for sustainable environmental management (Jodha, 1998). Similarly, an aware and motivated civil society can play an active role in environmental conservation.

Lapses from sustainable human development, leading to the secular deterioration in Pakistan's environmental health, are evident from several key indicators. Roughly 38 percent of Pakistan's irrigated land is waterlogged and around 14 percent is saline. The annual rate of deforestation ranges from 2.5 to three percent. Carbon dioxide emissions are increasing annually at a rate of 8-10 percent. And an estimated 250 million gallons of untreated water out of Karachi is dumped into the Arabian Sea every day, causing great harm to both humans and the ecology. Water toxicity, triggered by the rural-urban interface, has begun to reach alarming proportions. While not fully documented, its debilitating health effects have begun to manifest themselves with increasing frequency, especially in the proximity of large cities and towns. Recent data indicates that over one million acres of fertile, arable land in the Indus delta has become saline and unusable. The retention of freshwater flows by large dams has allowed salt-water incursion on a large scale, a problem that is likely to be exacerbated by a rise in sea level induced by global warming (Khan, 1999).

Two attempts to quantify environmental losses raise serious concerns about the sustainability of economic growth. Using extremely conservative assumptions, World Bank economists Brandon and Hammond (1995) estimate that environmental degradation results in the loss of about four percent of GDP annually, thus considerably offsetting the conventionally measured GDP gains. The breakdown of the GDP loss includes annual health impacts of water pollution (759 million US dollars) and air pollution (301 million US dollars), and the loss of agricultural production due to land degradation (300 million US dollars). Another way of interpreting the same data is in terms of the nation's net savings rate, after accounting not only for the depreciation of physical capital, but also for the degradation and depreciation of 'natural capital', accounting for the impacts of energy and mineral resource depletion, net deforestation, and emissions. A low savings rate means a lack of adequate resources for investment and growth. The World Bank (1995) estimates the 'genuine savings rate' for Pakistan to be only 2.5 percent of its GDP.

The Poverty-Environment Nexus: The Indirect Route

While Pakistan has achieved impressive aggregate economic growth, with GDP growth? averaging over five percent over the past three decades and a steady increase in per capita income (currently assessed at 480 US dollars in real terms and at approximately 2,000 in purchasing power parity terms), its impact in terms of poverty reduction is more ambiguous. Poverty is a multidimensional term. 'Consumption poverty' refers to the extent to which the private consumption of individuals or households falls below the 'poverty line', the minimum acceptable standard of private consumption. Another important dimension focuses on 'human development', a term which encompasses improvements in education, health, water, and sanitation in addition to the provision of sustainable livelihoods.

The proportion of Pakistan's population below the poverty line fell from 46 percent in 1984-85 to 34 percent in 1990-91 (World Bank, 1995), but has stagnated around this level in recent years. Less clear is whether the improvement in consumption poverty was accompanied by a reduction in the absolute number of people falling below the poverty line. Pakistan did not perform well by another measure of poverty, namely income inequality, as represented by the income share ratio and the Gini coefficient. Both have worsened over time. Finally, Pakistan's performance is even less creditable with regard to human development. Its social indicators are far below those of low-income countries with comparable or lower levels of per capita income. Furthermore, disparities across provinces as well as across the rural-urban divide are extremely pronounced (Khan, 1999).

A cyclical, downward spiral defines the relationship between poverty and environmental degradation. In the first place, poverty increases the vulnerability of the poor to degradation. Second, by restricting choices and entitlements for the poor, it turns them into potential predators of natural resources. However, the converse is that limited choices also create a reason to nurture resources and use them in a sustainable manner. A credible construct that contextualizes the poverty-degradation relationship is probably a blend of the two. In other words, poverty does not necessarily induce degradation, but if it happens to, it is a reflection of natural responses rather than deliberate acts.

Impacts of Degradation on the Poor: Resource Capture and Ecological Marginalization

Development that empowers the few, population growth, and the growing scarcity of natural resources combine to create a situation known as 'resource capture'.

> Resource capture occurs when population growth combines with a decline in the quantity and quality of renewable resources and the spread of market incentives to encourage powerful groups to alter the distribution of resources in their favor. Resources are, in effect, appropriated by elites, increasing environmental scarcity among poorer or weaker groups as a result. The manner in which this is done is through conversion of land from customary tenure to formal land titling. Groups experiencing this scarcity are then often ecologically marginalized as they migrate to rural or urban areas that are ecologically fragile. The resulting high population densities in the receiving areas, along with the migrants' lack of capital and knowledge of how to protect local resources, act to generate further environmental damage and chronic poverty. As scarcities of resources such as forests and urban land worsen their prices increase which leads to more acquisitive behavior, in effect creating a self-perpetuating cycle. (Gizewski and Homer-Dixon, 1998, pp. 9-10).

One of the key consequences of resource capture is ecological marginalization, where appropriation forces entire communities to migrate to marginal areas. Population growth compounds the problem. It is an outcome of social sector neglect and a factor that, subsequently, makes its way into the benefits of growth. Ecological marginalization exhibits itself in a number of ways.

One aspect is accelerated intra-rural migration. In the NWFP in Pakistan, for example, land-starved rural populations have moved into marginal lands, including erosion-prone hill areas and fragile semi-arid areas that have traditionally served pastoral groups and their herds. This has resulted in conflicts over land access. Consequently, low-quality pasture is now used more intensively and pastoral herds, forced to graze on post-harvest crop residues, are not fertilizing the agricultural land as much as before. As resources come under pressure from increasing numbers of land-poor and landless rural people, traditional management, tenure, and rights systems face collapse. Thus, grazing rights are being encroached upon and tree rights are subsumed under land rights, leading to deforestation and land clearing. State lands are under constant threat from farmers, loggers, and fuelwood collectors. The refugee influx from war-torn Afghanistan next door has only aggravated the problem.

A different and more complex situation prevails in the northern areas of the country. In these areas, the resource management problem is a function of cognitive space, property regimes, and scarcity (EDC, 1992). Exposure to market forces has expanded the economic horizons of the remote mountain valley dwellers, providing an outlet for an expanding population and commensurately reducing pressure on the environment. A consequence of this is that "degradation of their surroundings is less of a concern to the villagers because its importance is not as dominant as it used to be" (EDC, 1992). Also, migration to the small towns that have sprung up along the KKH has given rise to new problems of urban pollution. In the more inaccessible valleys, the people remain more sensitive to degradation problems and traditions of communal participatory management remain strong. Meanwhile, there is also greater receptiveness to community-based, participatory interventions that build upon such practices.

Sector Impacts and Vulnerability

This section analyzes the vulnerability of marginalized populations at the sector level. The effects on the poor are especially destabilizing because of the limited number of choices available to them. Consequently, the potential for social instability and conflict is high.

Water Pollution

Water pollution has three main sources: bacterial and organic liquids and solids from urban and rural domestic sewage; toxic metals, organic pollutants, acids, and other toxic substances from industrial discharges; and chemical pollution in the form of pesticide and fertilizer runoff from agricultural lands (Bakhtiar, 1992). Although the pollution caused by organic and chemical discharges is widespread, it predominantly affects the poor as a result of the skewed distribution of sewage, sanitation, and piped water facilities, both across the rural-urban divide and in poor urban slums.

In the older parts of cities sewage infrastructures are both poorly designed and poorly maintained, often resulting in the frequent mixing of raw sewage with drinking water. Thus, even piped drinking water in poor urban localities can be highly contaminated. Where it is not available, residents are compelled to use groundwater accessed through dug wells with shallow hand pumps. This water also contains bacterial and chemical impurities, as does the water in streams and rivers, which is also often used for drinking and washing. Dumped close to water sources, disease-causing pathogenic substances seep into the ground and into the water. Also, the poor are often forced to settle in low-lying urban areas, which although relatively more affordable are many times more prone to flooding, water pollution and the spread of disease.

In rural areas, the poor depend on rainwater that accumulates in ponds, rivers and springs, as well as on groundwater extracted from dug wells using hand pumps. The absence of modern sanitation practices and prevalent use practices results in contamination of the water supply. In the periphery of urban areas, where rural migrants tend to concentrate, runoff from nearby industrial and residential areas also contributes to water pollution.

The problem is compounded by the fact that there is a chronic lack of health facilities to deal with the effects of water pollution. Absence of adequate nutrition, lack of education and overcrowded housing increase human vulnerability to disease. A telling statistic is that infant mortality continues to remain high even though most demographic indicators have improved, a result of infants' high rate of exposure to waterborne diseases. The better off are better able to fend for themselves, given their access to household alternatives such as periodic cleaning and lining of underground water tanks, water filters and mineral water. The poor are left to bear the brunt of the health costs of water pollution.

Air Pollution

Air pollution ranks just as high as water pollution in terms of health and other effects on the poor. Past and projected trends in air pollution are linked closely with the country's energy profile, as well as its use efficiency. Two factors are of concern from a poverty perspective. First, the reliance on biomass continues to be high, close to 40 percent of the total energy consumption (Khan, 1997). While it is predominantly used in rural areas, biomass is also an important fuel source in urban slums. Second, among modern fuels, oil is rapidly replacing natural gas as a source for energy generation and industrial consumption. Moreover, as gas supplies dwindle, substitution with indigenous coal is becoming increasingly feasible. The use of both coal- and oil-based fuels in industry, energy, and transport sectors has had, and will continue to have, adverse health implications for the poor, especially in and near urban areas. In particular, the use of subsidized diesel fuel in public transportation results in extremely high levels of toxic emissions.

Traditionally, non-biomass-based air pollution was considered to be an urban phenomenon. More recently, the location of industrial sites, the growth of brick kilns and the penetration of transport into rural areas has made air pollution a rural problem as well, particularly in peri-urban areas and along major national highways. For instance, it would be difficult to find an uninhabited stretch of more than five miles along the Grand Trunk Road between Lahore and Islamabad.

Low-income neighborhoods spring up around industrial areas and power plants, where exposure to air pollution is high. The poor also work long hours in factories under unsafe conditions and women and children are particularly vulnerable, exposed to dust and chemical inhalations in sweatshops and industries that produce textile items, carpets, and leather goods. Traffic congestion and the resultant vehicular emissions are becoming an increasingly serious problem in Pakistan's big cities. Poor communities are the most exposed to automobile and other toxic fumes, as they tend to live close to the main trunk roads.

From a health perspective, the incidence of respiratory diseases and lead poisoning (predominantly among children) from mobile and stationary source emissions is escalating rapidly. As in the case of water pollution, factors that increase the vulnerability of the poor include poor nutritional intake, crowded living conditions (increasing the risk of exposure), and lack of access to medical facilities. Trapped in a vicious cycle, those who are the most disadvantaged are ultimately made even more so.

Chronic respiratory infections are caused by exposure to biomass emissions from cooking and heating in confined spaces, in rural homes and urban slum households. Women suffer much higher health risks as they do the cooking exclusively, and children and infants are equally exposed since they stay close to their mothers. In rural areas, poor, uneducated farmers are exposed to the detrimental effects of agricultural chemicals. Excessive use, drifting sprays, leaky applicators, and a lack of knowledge of how to handle dangerous substances all serve to enhance health risks. Similarly, women engage extensively in cotton picking, which consumes more than 90 percent of the total fungicides and pesticides used. According to a 1991 sample survey of 88 cotton pickers (mostly women), only one percent of them were safe from the hazardous effects of pesticides. Seventy-four percent had blood acetlcholine esterase (AchE) inhibition and between 25 and 50 percent were in a dangerous condition where blood AchE inhibition was between 50-87.5 percent (SDPI, 1995).

Solid Waste Management

Developing countries in particular, are ill-prepared to deal with the problem of solid waste, whether it involves dumping by foreign companies or it is internally based in rapid urbanization and industrialization. Once again, the burden falls disproportionately on the poor, a consequence of distorted municipal budgets and planning priorities.

In most cities in the south of Pakistan, between a third and a half of all solid waste produced in urban areas remains uncollected, often accumulating on roadsides and streets. In Pakistan, an average of 50,000 metric tons of waste is generated every day. Municipalities collect only 60 percent of this waste. Also, of the 8,000 tons or so of excrement produced daily, about half is deposited by the roadside or incorporated into solid waste. The waste that is collected is dumped in the open or, more frequently, burnt. Commercial, industrial, and hospital waste contains a high percentage of 'hazardous' or 'toxic' substances. These are often disposed of, untreated, as normal liquid or solid waste into drains, bodies of water, or open dumps without any special precautions. In general, there is a lack of space for waste collection bins and dumping sites. Municipal authorities lack human resources and vehicles for waste collection and disposal, and they make little or no effort to instill a sense of civic duty in the neighborhoods they serve.

Not only are waste disposal systems inadequate, they rarely serve low-income settlements at all. In fact, the solid waste collected in upper- or middle-class areas of cities is often dumped in the slums and city peripheries, thrown either into landfills or directly into watercourses. The poor urban areas are generally the worst served by garbage collection services, if they are served at all. The resulting problems are smell, disease-causing bacteria, and pests attracted by garbage and overflowing, clogged drainage channels. The ill-planned agglomeration of squatter settlements makes waste collection a difficult task, even when amenities are made available. Often, the dense housing and narrow winding streets do not allow vehicles to pass through, and with unaffordable operation and management costs (since the government sets inflexible standards), structures tend to deteriorate rapidly.

Not surprisingly, this has an enormous number of health implications. The untreated waste is loosely covered with soil and tends to absorb moisture, contaminating the underground water that the poor use for drinking, cooking, and washing. The resulting health impacts have been noted in the earlier section on water pollution. Municipal waste also contains leftover food, which attracts animals and insects. These are another source of communicable disease, affecting scavengers who pick recyclable and reusable waste out of the dumps. As these people live in congested (usually one-room) houses, the diseases are easily passed on to other members of the family. Medical waste and toxic waste from factories, which is not treated separately during collection and disposal, is another carrier of disease. Residual waste that scavengers are unable to sell (such as torn plastic bags, pieces of tires, and pieces of cloth) is used as fuel, producing harmful emissions. Burning, a common method of getting rid of accumulated waste, also produces toxic gaseous emissions which affect poor people living in close proximity.

Deforestation

Historical pressure on natural resources and ambiguously-defined property rights, overlain with rigid and increasingly corrupt management, has contributed to rapid deforestation in Pakistan. Even if we defer to the official version that forest stocks are increasing, it is a doubtful achievement. In the first place, the increase is based on an extremely low number as only four to five percent of Pakistan's land is presently under forest cover. Further, the increase conceals diverging trends in forest types. In other words, primary forest cover is declining, precipitating losses in genetic diversity and resilience.

Independent estimates, however, show wood biomass disappearing at a rate between four and six percent per year; it is feared that this number is the second highest in the world (BAP, 1998). It is estimated that if the present trend continues, Pakistan's total wood biomass could be consumed totally within the next 10 to 15 years (BAP, 1998). Indeed, Pakistan is one of the developing countries with no remaining undisturbed forests (WRI, 1997). A less-controversial finding is that both the quality and composition of forest stands is deteriorating. A recent study shows that good quality, tall forests (with more than 50 percent cover) occupy less than 400,000 hectares of land in Pakistan (BAP, 1998). Data collected through satellite imagery presents an even gloomier picture, showing that only 308,000 hectares remain under dense forest cover (PFRI, 2000).

The root cause of degradation lies in forest management practices, which have focused more on economic than environmental utility. Such practices are also to the detriment of community subsistence needs. Colonial governments had originally weakened community rights to the use of forest resources. Usufruct rights remained but were heavily proscribed. Community management traditions, already fragile, have further eroded with new opportunities for employment and emigration. Also, demographic and development pressures have forced communities out of their ancestral lands into marginal areas, where competition for resources is severe, resulting in further violations of indigenous property rights. The situation contains the seeds of conflict, with communities forced to act as predators rather than as guardians of the commons.

Weak property rights and increased fears of expropriation by powerful groups fuel insecurity about future earning potential, especially in an inflationary environment. In general, this gives rise to concerns about the free rider problem, namely, that forests will disappear no matter what the community might do (Inayatullah, 1996). Even in privately-owned forests (*guzaras*), "right holders may see in regeneration a reintroduction of state property rights, which may stifle even natural regeneration" (Azhar, 1993).

While there is little doubt that under hostile management and tenure regimes, communities are showing a propensity to raid forest resources, their activities pale in comparison with the activities of the so-called 'timber mafia'. These are commercial loggers who, driven by rising timber prices, are willing to undertake illegal logging. The timber trade also demonstrates a distinct anti-community bias, as while communities are entitled to a substantial share of royalties from logging in *guzara* forests, active collusion between the 'mafia' and the forest department results in the appropriation of the bulk of these royalties.

Agricultural Land Degradation

The 'green revolution' strategy of agricultural development was based upon a sufficiency of both land and water. Major irrigation schemes (fed by large dams built at Mangla and Tarbela) were launched to harness this water and to apply it extensively. The intensive application of water and chemicals to high-yielding dwarf varieties enhanced crop yields in wheat, rice, cotton, and maize to a remarkable degree. However, questions have been raised regarding the longer-term viability of this strategy. Evidence shows that the country's agricultural resource base, which is made up of both land and water, has been unsustainably mined. The drainage system has not been able to cope with the change in irrigation practices, creating serious problems of waterlogging and salinity.

Policies and public sector management systems have given little space to incentives as a means of conserving the above resources or using them efficiently. The economic aspects of this strategy, focusing on output price supports (which have kept prices of crops such as cotton and wheat below border prices) and compensating input subsidies, have contributed to environmental degradation in a number of ways. The price support system has created relative price differentials, depressing the prices of wheat and cotton and increasing the price of sugarcane above world price levels, leading to over-production of this highly water-dependant crop.

Environmental impacts stemming from input subsidies are even more pronounced and wide-ranging. Input subsidies include those on irrigation, electricity, pesticides, fertilizer, seeds, and machinery. Excessive use of fertilizers and pesticides has given rise to water pollution and soil degradation. A good review of these effects can be found in Khan (1999). A combination of low water charges and inadequate operation and management has contributed to severe waterlogging and salinity problems.

Adverse terms of trade for food and cash crops have led to inter-sectoral transfers of income and to the further impoverishment of the rural poor. Small and landless farmers are most affected, with their indigence leading to increased pressure on natural resources. Cropland degradation affects the health and nutrition status of the poor and lowers their productivity. This failure can happen directly through lower yields per unit of labor or land because of reduced soil quality. It can also happen indirectly, through the reduced physical capacity of laborers because of malnutrition and poor health. Even in cases where the poor are healthy, productivity can be low due to increased time allocated to less productive activities, such as fuelwood collection. As a result of deforestation, family members have to go further to collect fuel, which takes time away from farming activities. Also, as fuelwood becomes scarcer, animal dung is substituted as fuel, resulting in non-replenishment of soil nutrition and, eventually, soil exhaustion.

Case Studies

The previous sections have established the link between environment and security as a complex and multifaceted relationship, defined environmental security as a component of larger human security concerns, highlighted the multiple cause-and-effect relationships between and among different sectors. In studying the subject, focus should be on the impact on the poor and marginalized since they are disproportionately affected and least able to cope with new layers of vulnerabilities added via environmental insecurity.

To further elaborate on these points, this section will present two case studies. These studies are selected on the basis of two criteria. Firstly, they illustrate how each particular episode tends to contain its own particular route to conflict, and secondly they focus on equity at a national and local level. The first is a detailed case study of a national water policy issue in Pakistan: the Kalabagh Dam. The second focuses on a local forestry-related episode where outright conflict has erupted.

The Kalabagh Dam Controversy

The Indus River, the agricultural lifeline of Pakistan, flows southwest for about 2,500 kilometers and empties through an immense delta into the Arabian Sea. The river system and its tributaries have provided Pakistan with some of the most fertile and best-irrigated land in the subcontinent, about 200,000 square miles' worth. Two large dams across the Indus river (the Tarbela and the Mangla) and a vast supporting irrigation network have both fragmented and altered its original ecosystems. A growing concern is that yet another dam across the Indus would prove environmentally disastrous, and that it could trigger irreversible degradation of an already fragile ecosystem.

The expectation is that the Kalabagh Dam would store an additional six million acre feet (MAF) of water. Also, it would produce over 1,200 megawatts of hydroelectric power. However, the global and regional context for assessing large dams like the Kalabagh is changing, with conventional irrigation, flood control, and energy benefits being viewed through the prism of sustainable development. This provides the contextual backdrop for a review of the proclaimed benefits.

Water Availability

In the first place, the water availability claims used to justify the construction of a major dam at Kalabagh are dubious. The Water and Power Development Authority (WAPDA) has used a wet cycle period to show higher inflows at the rim stations (146 MAF). The longer period, which includes both wet and dry cycles, indicates relatively lower inflows (139 MAF). In addition, in order to show the excess that would be available for use, spurious inflows have been shown on the eastern rivers, rim station amounts are inflated to add to downstream flows, water losses are understated, and diversions to India under the Indus Basin Accord are not factored in. If the Kalabagh Dam were to be built on these false premises, the associated irrigation infrastructure would mean that the upstream province of Punjab would divert even more water from the already water-starved downstream province of Sindh.

Waterlogging and Salinity

As we have already discussed, waterlogging and salinity are among the more severe environmental problems facing Pakistan, and their economic ramifications are almost incalculable. Building the Kalabagh Dam would mean even higher water retention in a system ill-designed to cope even with existing inflows.

In fact, the waterlogging and salinity problem is so acute that a controversial and exorbitant (780 million US dollars), 25-year-long National Drainage Plan has been launched to mitigate its impacts. The Kalabagh Dam would surely add to the problem, not only in its immediate environs but also where new irrigation infrastructures would be located.

Table 3.2
Land, in million hectares, with water table depths under five feet (by province in the June-October period)

	Punjab	Sindh	NWFP	Balochistan	Pakistan
1988	0.54	0.86	0.06	0.04	1.50
	1.72	3.44	0.06	0.09	5.31
1990	0.71	2.34	0.05	0.09	3.20
1992	0.64	2.23	0.05	0.14	3.05
	1.25	4.08	0.06	0.14	5.53
1993	0.58	1.30	0.04	0.07	1.99
	0.98	4.60	0.06	0.15	5.25

Source: *Compendium of Environmental Statistics of Pakistan, 1994-95* (GoP, 1998)

Table 3.3
Land, in thousand hectares, affected by salinity (by province)

	NWFP	Punjab	Sindh	Total Indus Basin
Canal Command Area (CCA)	320	7,891	5,351	13,562
Affected Area within CCA (Percent)	14 (4.3%)	1,614 (20.4%)	1,532 (28.6%)	3,160 (23.3%)
Affected Area outside CCA	502	1,129	1,019	2,650
Total Affected Area	516	2,743	2,551	5,810

Source: *Soil Survey of Pakistan, 1977-78* (GoP, 1980)

As Table 3.2 shows, the province of Sindh is already worst hit by waterlogging. Incremental land degradation is likely to be most pronounced there, thanks to the Kalabagh. Table 3.3 indicates that incidence of salinity is also higher in Sindh than in Punjab. Clearly, the Kalabagh Dam would add yet another wrinkle to an already fragile political relationship between the two provinces.

Although the information in Tables 3.2 and 3.3 is quite dated and recent data is not available, the numbers are nonetheless important and relevant today. For example, Table 3.3 shows that even twenty years ago almost 30 percent of the area within the canal commands in Sindh were afflicted by salinity, as compared to 20 percent for the Punjab. There is a general consensus that the situation has only become worse over the last two decades, arguably more so for Sindh.

The upshot is that attempts to increase crop production by tapping new sources of water could be self-defeating, thanks to the soil degradation resulting from it. There is an institutional dimension to this as well. Large farmers are liable to appropriate the bulk of the additional water under the existing supply-based distribution (*warabandi*) system.

A clearly preferable choice is to use existing water more efficiently and to focus on making the necessary institutional changes for its equitable distribution. Some of the proposed measures include the rehabilitation of select canals and watercourses (in saline groundwater zones), commercial water distribution, re-pricing of water to reflect adverse environmental externalities, and decentralized and participatory water management. These are clearly win-win solutions, since they are relatively low-cost, efficient, equitable, and environmentally-friendly.

As a rule, however, vested interests tend to prevail when there are expectations of reaping construction or irrigation benefits. Such benefits are garnered within the framework of loose and inequitable governance, at considerable cost to the national economy, the people, and the environment. Thus, flexibility of policy and institutions becomes imperative under the current circumstances.

Ecosystem Degradation

The degradation of the Indus delta ecosystem, as a result of reduced water outflows, is already a highly visible phenomenon. The present level of silt discharge, estimated at 100 million tons per year, is four times less than the original level before the rivers were dammed. The combination of salt water intrusion (some reports show this as far as 30 kilometers inland), and reduced silt and nutrient flows has changed the geomorphology of the delta considerably. The area of active growth of the delta has fallen from an original estimate of 2,600 square kilometers (growing at 34 meters per year) to about 260 square kilometers. Fresh water reaches only a few of the creeks and others have become blocked. The delta is being transformed by strong wave erosion, an increasing dominance of sand at the delta front, and an increase in wind-blown sand deposits resulting from losses in vegetation.

The consequent ravages to Pakistan's ecosystem have been exceptionally severe, in particular to the mangroves which are its mainstay. They sustain its fisheries, act as natural barriers against sea and storm surges, keep bank erosion in check, and are a source of fuelwood, timber, fodder and forest products, as well as a refuge for wildlife, and a potential source of tourism. Without mangroves, the nutrients they recycle, and the protection they provide, other components of the ecosystem would not survive. The direct and indirect benefits of mangroves are enormous. In 1988, Pakistan earned 2.24 billion Rupees from fish exports (around 100 million US dollars), 72 percent of which was made up of shrimp and prawns. The collective income earned from fuelwood, fodder, and forest products was another 100 million Rupees (around 4 million US dollars). These are orders of broad magnitude that are threatened by mangrove degradation. Even where official numbers are absent, the functions are, in themselves, indicative. For instance, substituting natural barriers with physical ones, like dykes, walls, and dredgers, would require enormously high capital and maintenance costs. In addition, the wildlife and tourism potential of the mangrove swamps has not yet been exploited.

The condition of the mangroves is directly linked to fresh water outflows. Releases below Kotri barrage (in Sindh) average 34 million acre feet (MAF). Of this, about 20 MAF actually reaches the mangroves, and that falls between the rain-fed (*kharif*) months of July and September. The rest is lost due to evaporation or diversions. According to the Sindh Forestry Department, about 27 MAF are required to maintain the existing 260,000 hectares of mangroves in reasonably healthy condition. This is seven MAF more than is currently available, a situation that has contributed to significant ecosystem instability and mangrove loss. Within the framework of the Indus Water Accord, the intent is to divert an additional 11 MAF for upstream dam construction, including the Kalabagh Dam, in order to meet agricultural and hydropower needs. This would further reduce already sub-optimal flows and aggravate an already dire situation.

About 100,000 people live in coastal villages on the northern side of the Indus Delta. The mangroves are a vital source of livelihood for them, both direct, in terms of fuel, fodder, and grazing, and indirect, in terms of fish and amenity values. The prevailing view is that since they are poor, these communities are prone to degrade their environment. However, it is difficult to fathom why poor communities would choose to endanger the very basis of their existence. The more likely explanation is that their practices have not changed, but they appear unsustainable because the resource base has begun to degrade. Communities are more often the victims than the agents of such degradation. The real culprits are water diversion, biological and chemical water contamination, and large-scale commercial practices, compounded by institutional ignorance and complicity.

Degradation of the Indus Delta ecosystem is not the only manifestation of the 'biodiversity deficits' that are emerging along the entire length of the Indus River. The ecosystem has been severely fragmented over time by its extensive network of canals, dams, and barrages, resulting in threats to a variety of species and organisms. For instance, two species facing extinction in the lower reaches of the Indus River are the Indus dolphin and the *palla* fish. Both can be classified as indicator species, and their impending extinction represents the loss of a way of life that was characterized by interdependence between communities and their environment. To all intents and purposes, most of the river Indus has been nationalized. This has led to the denial of fishing rights to river-based communities and the wanton exploitation of river resources through contractual arrangements. The new eco-equations, habitats, and planned captive breeding programs are poor substitutes for natural genetic resilience.

The Crisis of Governance

The Kalabagh Dam is a classic case of environmental insecurity, which has taken a turn for the worse. It typifies government insensitivity, in this particular case, to inter-provincial concerns regarding water availability, environmental degradation, and social displacement. An important factor in good governance is decentralized and consultative decision-making. By contrast, Kalabagh has been the very antithesis of this, with policy decisions made in a highly centralized, often secretive, politically coercive, and technically flawed manner. Indeed, WAPDA, the government agency responsible for large water projects, appears intent upon an ex-post vindication of a politically-motivated step.

Regrettably, when there is a need for broad-based stakeholder consultations as a basis for informed and democratic decision-making, the existing trend points towards even greater centralization. For instance, the rotating chairmanship of the Indus River System Authority has recently been converted into a permanent appointment, provincial resolutions against the Kalabagh Dam project have been given short shrift, the Council of Common Interests has consistently ignored the matter, and community concerns continue to be met with blatant disregard. The outcome is that civil society and political leadership in the smaller provinces are up in arms against the Kalabagh Dam, while national policymakers remain intent on making it a reality.

The present drought situation typifies this alarming blend of arrogance and ignorance. In defiance of all beliefs, WAPDA continues to lobby for the dam even as the Indus runs completely dry in the most improbable place: the confluence of the four main feeder rivers. Suggestions as outré as laser-melting the glaciers are gaining currency in the popular press and matte-finished, ageless generals herald 'Comprehensive 50-Year Water Master Plans' with large dams as their focus. More substantively, Sindh continues to take proportionate reductions in water, with the result that river flows, from downstream of Kotri all the way to the sea, have stopped completely.

In summation, the anthropogenic activity that degrades the Indus Basin's ecological integrity is an ongoing process and a culmination of many factors. It is instigated and sustained by developmental and demographic pressures. It reflects the imposed dominance of technology over nature. It is an outcome of weak institutions succumbing to vested interests. It represents a failure to exploit the traditional synergies, practices, and interactions between communities and their environments. In the final analysis, the aforementioned degradation is a symptom of the piecemeal and extractive manner by which natural resources are utilized. This contrasts with the common sense embodied in Abramovitz's (1996) remark that "freshwater ecosystems are the critical link between land and sea, in effect forming the planet's circulatory system; virtually every human action is eventually reflected in them." This holistic view, which integrates spatial, biophysical, and human dimensions, should form the core of all efforts to manage ecosystem resources sustainably.

In the Kalabagh case, feudal interests on both sides have proven to be moderating factors, preventing outright conflict between the provinces. While the institutional arrangements are vulnerable and are constantly being manipulated, enough political pressure is being placed on both sides to ensure that the burden of reduced water availability remains equally distributed. But it is an uneasy compromise, maintained by a shaky military government, which can ill afford another outbreak of inter-provincial discord. For the moment, this consideration has kept even WAPDA at bay. However, when the antagonists fall into different social zones the scope for mediation is far more limited, as demonstrated in the case study below.

Deforestation: Communities and Contractors

Forests have many important economic uses and are a source of livelihood for many communities. In addition, many ecological and environmental imperatives are associated with them. They are a source of biodiversity and, as watersheds, they regulate water and sediment flows downstream. Given their many uses, the need for sustainable forest management becomes an imperative. Unfortunately, the term is more honored in its breach. The data shows a rapid decline in both coverage and the quality of stands. Also, forest degradation has led to a spate of onsite and downstream ravages such as biodiversity loss, land erosion, flooding and dam sedimentation.

The management system, designed for a specific purpose, has been unable to cope with these changes. The multiple and often conflicting interests of commercial loggers, private developers, governmental and military agencies, hunters, and impoverished communities has placed this system under relentless strain. The forest department tends to choose the path of least resistance, coming down with a heavy hand on the disempowered communities and colluding for personal profit with vested interests. Officials have become increasingly vulnerable to outside economic inducements, as opportunities for financial and professional betterment become hostage to fiscal insolvency.

Many factors combine to create a complex web of perverse incentives antithetical to conservation: rising prices of timber, fuelwood, and forest products; an erosion in the living standards of the forest custodians; fines and penalties that are applied selectively and fail to match the nature of transgressions; and royalties that are appropriated by the rich and powerful. The irony is that the commercial and development groups, which management is not in a position to oppose, are making the key inroads into forest resources and in fact cooperate with the transgressors. On the other hand, it has targeted communities whose needs are of a subsistence nature and who, if their rights and traditions are properly protected, can work harmoniously with the authorities toward the sustainable management of forests. Box 3.1 provides an example of how the royalty system, ostensibly designed for the fair distribution of forest benefits, is manipulated by the timber mafia to their own advantage.

Box 3.1

Forest royalties in the Northwest Frontier Province

(Based on Knudsen, 1995)

Forest royalties lie at the core of the problem of perverse incentives. With the enormous financial stakes involved, most of the revenues tend to get skimmed off before reaching the communities. The system is not only inequitable, but is also designed to induce deforestation rather than conservation. In both protected and *guzara* forests, where communities have legal rights to forest resources, the net revenues (after deducting costs) from timber sales are divided between communities (in the form of royalties) and the government. The distribution is as follows:

- Prior to 1981, the fixed-price system was adopted, where the concessionaires were paid a fixed price per cubic foot of the total harvested volume. This system did not keep pace with the rising price of timber.
- Subsequently, the system was replaced by the net-sale system. Under this system, the harvested timber is auctioned at timber markets. The net proceeds are then divided between communities and the government, with the community's share ranging from 60 to 80 percent. Ideally, this allows communities to receive the benefits of rising prices.

In effect, the timber contractors, elected members, and designated community representatives manipulate the system for profit in collusion with the forest department. Common malpractices include:

- After being awarded a logging contract, the contractor/owner delays logging deliberately in the specified block, in anticipation of rising prices because he has the financial means to engage the forest department in extended lawsuits. He meets the existing demand from his own private stands.
- Using the clout of local notables, timber contractors obtain the power of attorney for village concessionaires, enabling them to collect royalties from the government. These royalties are retained after payment of a pre-agreed amount to the local notables and forest department staff.
- Under the net-sale system, timber contractors purchase royalties from local communities for a fixed price slightly above the old fixed rates (50 Rupees per cubic foot). Communities are willing to forego higher but delayed payments for quicker up-front disbursements. The contractor then bids on the contract under an assumed name. He has

an incentive to log more than the agreed volume (facilitated by bribing forestry department staff), as this means more royalties for him.

While, admittedly, these malpractices are possible due to loopholes in legislation, the politically powerful contractor networks not only enjoy immunity from the law, but they also resist attempts at legal reforms which would close such loopholes. An alternative system (Inayatullah, 1996) "involves a change in the formula of payment from royalties to income; in other words, owners receive an annual payment based on the number of trees standing in their area, not on the number of trees sold." Implementation of such a system pre-supposes extensive documentation and information about existing stands.

Such malpractices are prevalent in both the NWFP and in the northern areas of Pakistan. In Dir-Kohistan, the appropriation of royalties has led to outright conflict. Interestingly, even recourse to the legal system has produced no reprieve, even when decisions ostensibly favored communities. Dir Kohistan is a valley of the Panjkora river, which is more than 70 kilometers long. Kalkot and Lamotai are villages located in the upper portion of that valley. At the end of the 17th century, the British government adopted a policy of indirect rule in the area, delegating local *nawabs* to govern on their behalf and collect revenues (*maaliya*). Reciprocally, the British turned a blind eye to the *nawabs*' autocratic rule. However, communities were usually allowed to meet their subsistence needs from the forests.

After independence, Dir Kohistan was merged with Pakistan, coming under the direct authority of the Pakistani government. Under the new dispensation, the rich forests were declared state property and the communities there were promised 15 percent of the income from the forests as a royalty. Commercial logging began on a large scale. Initially, this was done through contractors who cut more trees than the legal limit, did not share the proceeds, and went so far as to disallow communities their traditional subsistence rights. Discord began as early as the 1970s, erupting into outright violence when authorities resorted to force and shot dead a number of community leaders.

The then Prime Minister, Zulfiqar Ali Bhutto, visited the area and abolished the contractor system, replacing it with the Forest Development Corporation (FDC). The community's share of the royalties was increased to 60 percent. This could be an enormous windfall, provided the communities received their due share. As box 3.1 shows, very little actually filtered down to them.

Growing resentment against the government, the FDC and the contractors finally came to a head in the early 1990s. Village youth banded together to form the Kalkot Youth Welfare Society (KYWS). They set up a manned checkpoint to stop all movement of timber outside the valley, and both the written and spoken rhetoric became confrontational. The government initially attempted to diffuse the situation by setting up an Inquiry Committee in 1997 to investigate the community's grievances. The findings of the committee vindicated the community's stance and advised redress but the district administration failed to act upon its recommendations. Encouraged by its moral victory, the KYWS took its resistance to a new level. A smuggler was shot dead at the checkpoint by the community guards and new checkpoints were set up. The society organized peaceful marches and sit-ins, first locally in Shringal, Dir, and Timergarah and eventually in the provincial capital, Peshawer. However, no resolution of the problem is in sight. The provincial and district governments have adopted a hostile stance, accusing the communities of taking the law into their own hands. An uneasy stalemate prevails at present, with a real risk that the situation will soon escalate into a large-scale conflict.

Conclusion

Environmental security is a subset of human security. Its essence is that human beings, by virtue of their numbers and the magnitude of their activities, are causing bio-geochemical changes in the earth's system that are taking place many times more rapidly than those occurring naturally. The term 'activities' can be viewed as a catch phrase for unsustainable development processes. Whether they take the form of excess consumption, as in the north, or inequitable and resource-depleting outcomes, as in the south, these activities always contribute to environmental degradation and pollution.

The above case studies illustrate the insecurity-conflict nexus. An interesting paradox is presented, reflecting the interplay of social and economic forces. In the Dir case on forestry these forces have established an almost surgical divide between the antagonists. Specifically, communities are pitted against a consortium of contractors, the district administration, the forestry department, and 'community elders'. The latter act to subvert the movement from the inside. Institutional redress mechanisms, such as official investigations and judicial recourse, tend to lack transparency. Ultimately, they are subservient to powerful economic forces. There exists the very real threat that the combination of rapid deforestation and continued exploitation of the communities will escalate into large-scale violence between these actors.

In contrast, the Kalabagh Dam study presents a different dialectic. While the antagonists are still separated by economic and environmental interests, the lines of conflict are blurred by their social constitutions. Powerful lobbies exist on both sides of the divide, both with an interest in increasing water allocations. The government, as is its wont traditionally, defers to these lobbies, which results in an uneasy compromise. But the fragility of the compromise is underscored by the fact that it is driven by power rather than environmental or social logic.

Chapter Four

Environment & Security in Bangladesh

A. ATIQ RAHMAN
ZAHID H. CHOWDHURY
AHSAN U. AHMED

"Security means development.
Security is not military hardware though it may involve it.
Security is not military activity though it may encompass it.
Security is development
and without development there can be no security."
-- **Robert McNamara**

IN RECENT DECADES the concept of security has undergone significant changes, especially after the end of the Cold War. Relationships of power between states for military defense are no longer the pivotal points of security analysis. Instead, issues like socioeconomic and political development, as well as environmental and ecological factors that include such natural resources as soil, water, forests and fisheries, are becoming increasingly prominent in security studies (Ullman, 1983). Security issues have deviated from military concerns particularly in the developing countries, where the individual nation's capacity to deal with regular day-to-day problems is already strained by factors such as poverty, inequity, underdevelopment, malnutrition, decrease in per capita endowment, quality of natural resources, and undue pressure on the regeneration capacity of those, and so on (Homer-Dixon, 1991). Global issues such as the greenhouse effect, El Nino Southern Oscillation (ENSO) and others will only accentuate these issues and might further threaten the ecological sustainability and environmental security of vulnerable developing countries (Ahmad and Ahmed, 2000; Munasinghe, 2000).

Bangladesh represents a good case study of these changes since its vulnerabilities stem from both internal and external sources; however, most traditional security analysis has overemphasized the external aspects. More traditional constructs of security questions have now begun to be questioned (Islam, 1994; Iftekharuzzaman, 1994). For example, Iftekharuzzaman (1994) argues that, "societal impediments like the country's massive poverty and underdevelopment and illiteracy which, through their interface with the chronic political instability and crisis of governance, perpetuate the country's internal vulnerability." He also goes on to note that endogenous factors such as environmental threats, resource scarcity, and crop failure are often omitted or undermined, even though it would not be possible to ensure security without safeguarding the nation from internal vulnerabilities.

The aim of this paper is to link the security issues of Bangladesh with perspectives on the environment and resource management. We begin with an overview of the environmental situation in Bangladesh and then move to explore the links between environmental pressures and security concerns in the Bangladeshi context.

The State of Bangladesh's Environment

Bangladesh is currently home to about 130 million people. The population is growing, despite significant successes having been achieved over the last two decades in keeping the growth rate down, at a rate of 1.67 percent each year (Ministry of Finance, 1999). Its large population is putting immense pressure on the country's natural resources, which are already scarce and diminishing swiftly. Unfortunately, this is not the only factor triggering non-sustainable use of the natural resource base. Rather, it is a complex mixture of political, socio-behavioral, economic, and governance issues that is leading Bangladesh toward the depletion and degradation of its natural resources including water, soil, fisheries, forests and wetlands. Furthermore, unplanned urbanization and the unlawful release of industrial and household wastes have caused serious degradation in terms of air, water, and soil pollution throughout the country.

A significant percentage of Bangladesh's population is both poor and illiterate. As a result, their access to resources and information is extremely limited. Despite being poor, most citizens do care about environmental quality, especially at the grassroots level. It is partially rooted in their modest and candid approach to livelihood and behavioral practices. But the major cause for continued degradation of the natural resource base lies with failures at the governance and management levels, especially due to authoritative manipulation of legal provisions by local elites and the serious lack of implementation of legal provisions by government officials. In most cases the two problems go hand in hand.

Governance and Geopolitics

Bangladesh achieved independence in 1971 following a bloody war against Pakistan. However, its efforts to initiate a democratic regime had started even before it earned its independence. One of the major reasons for the movement towards independence was the failure of the ruling elites of what was then known as West Pakistan to share power with their Bengali counterparts. However, Bangladesh's journey under a parliamentary form of government did not last long after independence. The multi-party democracy faced a setback following a parliamentary coup in 1974, when a one-party system was initiated. The political situation of the country further deteriorated with a military coup in 1975. The military remained in control of power up until late

1990, often disguised as a pseudo-democracy, but never more than a military-controlled government.

A process of massive militarization of civil administration, politics, and society was pursued by the military regimes that have ruled the country for most of its existence. It almost completely destroyed the democratic norms, practices, and structures of the country. Even though the security threats to a developing country are mostly endogenous rather than exogenous, national security considerations of these military regimes in Bangladesh overwhelmingly favored external threats of military intervention (Maniruzzaman, 1992; Hafiz and Khan, 1990). The typical military outlook at the time was that "India has the scope and means to capitalize on Bangladesh's geo-politically locked situation" (Iftekharuzzaman, 1994). This definition of the security problematique naturally contributed to the expansion and maintenance of a large military force. Thus, the national security policy of the military regimes supported their own institutional interests. This can be noticed from the sharp increases in the defense budgets of Bangladesh from the mid-1970s onwards. In fact, the defense budget tells only part of the story. In order to deceive the people, a big portion of the military budget has been channeled through the allocations of other sectors, including health, education, and public works. If all these disguised figures were added up, Bangladesh's military spending would probably exceed one quarter of total spending.

Meanwhile, the Constitution of Bangladesh had undergone significant changes in order to facilitate the aforementioned security strategy. An artificial conflict of national identity has been developed through these constitutional changes, even though Bangladesh is traditionally a remarkably homogenous country. Its near homogeneity in ethnic, linguistic, cultural, and even religious terms makes it a unique case in the modern state system. Hence, it is noted that "unlike most other developing states of the Third World, which have been aptly described as nations not in being but only in hope, Bangladesh is a new state of an old nation" (Iftekharuzzaman, 1994). In spite of these unique advantages in cultural homogeneity, nation-building was not an issue for the ruling elite of Bangladesh in the post-independence era, and very little has been done as far as it is concerned. Rather, a pseudo-debate on whether ethnic-linguistic or territorial identity would be the basis of nationhood had become popular, therefore contributing to the political polarization of the people (Shelley, 1986).

Alarmingly, this debate over national identity continues to occur even after the transition to democracy, which followed a massive popular uprising in the late 1990s. The crisis of governance that had developed right from the nation's inception had deepened during several successive military regimes, along with the debate over nationalism, creating widespread confusion among the people as well as chaos and instability in the political system.

Such a chaotic and instable situation that created the crisis of governance was possible because of the absence of deep-rooted and efficient political institutions. Almost all the major political parties are organized on the basis of factional and parochial group interests rather than around any well-founded ideologies or a specific set of programs of action. Therefore, patron-client relationships pervade the whole system, creating a society devoid of democratic or institutional virtues. Such a situation automatically cripples the ability of the political elite to resolve issues such as poverty, unemployment, and other socioeconomic challenges. For the same reason, the non-military dimensions of security, such as poverty, underdevelopment, over-population, protracted conflict, resource scarcity, dwindling natural resources, and others, are not regarded as the primary concern of security strategists. This, despite the fact that failure to meet such socio-economic challenges can only serve to make the country even more vulnerable, may ultimately facilitate external interference in its internal affairs.

Physical Setting and Climate

Bangladesh lies on both sides of the Tropic of Cancer in the northeastern part of South Asia. It occupies a total area of only 147,570 square kilometers. The country is surrounded by India on three sides and shares 3,715.18 kilometers of common border with it; this is about 93 percent of Bangladesh's total borders. The only other contiguous neighbor is Myanmar, which shares a mountainous border in the southwest. The Bay of Bengal lies to the south of Bangladesh. The topography of the country's vast floodplains is extremely flat and over half of the entire landmass lies within 10 meters of sea level.

Like all tropical countries, Bangladesh has a hot and humid climate during most of the year with a mild winter between November and March. Being a small state, its climate is fairly uniform throughout the country. Despite the local calendar, which claims to have six seasons, three primary seasons are identifiable, including summer, monsoon, and winter. The hot, humid summer stretches from March to June. The rainy monsoon season, which is Bangladesh's most prominent season, lasts from June to October and is characterized by heavy, often torrential, rainfall. The winter is relatively cooler and drier. Annual rainfall ranges from 1,200 to 2,550 millimeters, on average. There are strong spatial and temporal variations in rainfall. The Eastern Zone has higher rainfall than the western zone and about 80 percent of the annual rainfall occurs in the five months between June and October. The evaporation of water in the winter appears to be greater than the amount of rainfall, and this unfortunately results in an acute shortage of topsoil moisture.

Demography

Although of a similar size to Illinois, Bangladesh's total population is about half that of the United States. About 50 percent of this population is below reproductive age (15 years old), meaning that it has a high growth potential. However, during the past two decades Bangladesh has successfully slowed its population growth rate, reducing it from three percent to approximately 1.67 percent. Remarkable achievements have also been made in terms of decreasing child mortality rates and increasing life expectancy. Despite such success stories, Bangladesh remains the most densely populated country in the world, with the exception of a few city-states. However, its population is dispersed evenly across the country except for the hilly southeastern districts.

A large proportion of the population is unemployed, while 63 percent of the employed labor force is predominantly engaged in subsistence agriculture. Unless labor-intensive industrialization becomes the major thrust of the economy in the near future, the nation will have to face a tremendous challenge in terms of providing employment to all. Lack of employment and other opportunities for income is likely to add to further social vulnerability in the coming decades.

Land

Bangladesh is located on the eastern portion of the Ganges-Brahmaputra-Meghna (GBM) delta, the largest in the world. It has a huge catchment area, covering 1.75 million square kilometers and crossing the borders of five countries. Formed by three mighty rivers, this is also the youngest and the most active delta in the world. It has been formed over millions of years from the sediments seeping down from the great Himalayas and Tibetan Plateau. The annual sediment load of the GBM river system is estimated to be 2.4 billion tons of cobbles, sand and silt (Rashid and Paul, 1987). Over 92 percent of the combined discharge of these rivers flows over the landmass of Bangladesh, which takes up only about seven percent of the total catchment area.

The deltaic floodplain of Bangladesh stretches northward to the foothills of Assam. It has a wide range of soil types: there are about 500 soil series ranging from juvenile new alluvium at one extreme to old red soils on the other, with various kinds of landforms and hydromorphic and drainage conditions in between (Rahman, 1994). Since Bangladesh is practically all flat alluvial plain, its land and soil conditions seem to be homogenous, but in reality its land pattern and soil condition are rather complex. The land systems in Bangladesh consist of three land types: floodplains, terraces, and hills. Each land system can be further divided based on zonal differences, topography, soils, drainage and hydrological conditions.

Floodplains: Around 80 percent of the total land area of Bangladesh consists of floodplains. Innumerable tributaries and distributaries of the combined GBM system criss-cross it. Within the broad floodplain landscape, some distinct landscapes are formed, based on environmental conditions under which the sediments were deposited. Six different types of these can be recognized: piedmont plain, basins, and active, meandering, tidal and estuarine floodplains.

Terraces: Terrace areas constitute only eight percent of the total land surface of Bangladesh. The terrace areas have been elevated 10 to 20 feet above the surrounding floodplains, including the Barind tract in the northwest and the Madhupur tract in the north-central region.

Hills: These areas include the northern and eastern hills and the Akhaura Terrace. Combined, they cover about 12 percent of the nation's total land area. Running north to south along the eastern border of Bangladesh, the northern and eastern hill ranges are linear ridges. The Akhaura Terrace is only 15 to 20 feet higher than the surrounding areas, while northern and eastern hills could be as high as 3,000 feet from sea level. Soils are very acidic and sandy in these areas.

Forests

Bangladesh once had very rich tropical forests. However, due to the fact that biomass energy is the major source of energy in rural areas, the country has been losing its forest resources at a very rapid rate, especially in the 1980s. Today, the major forest areas of Bangladesh include the Sunderbans, the Chittagong Hill Tracts, and various deciduous forests. These are further discussed below.

The Sundarbans: Located in the southwest, this mangrove forest covers an area of approximately 5,500 square kilometers. Together with the adjoining Indian side of the forest, the Sundarbans is the largest patch of productive mangrove forest in the world. This is the home to rich flora and fauna including the famous Royal Bengal tiger. Because of its rich biodiversity, the United Nations has declared the Sundarbans a World Heritage Site. Due to continued settlement and tree felling, the vegetative density has shrunk significantly over the years. Deterioration of the canopy cover and continued encroachment threaten the existence of the Royal Bengal tiger and a number of other species. The government is now contemplating a participatory restoration project with financial assistance from the Asian Development Bank (ADB) and the Global Environmental Fund (GEF). The future of this forest will largely depend on the success of the interventions proposed in the restoration project.

Chittagong Hill Tracts: These tropical rain forests are situated in the mountainous southeastern part of the country. The Chittagong tracts are home to rich vegetation and wild animals, including elephants. Their forests house several long-rotation, high-value timber species.

This valuable forest resource is now being rapidly depleted due to internal demand for timber wood and the growing energy demands of the local population.

Deciduous Forests: The only deciduous forest of the country, located in the high Modhupur tract, used to be covered with *sal* trees, which are an important commercial timber and fuelwood species. However, due to rapid human encroachment for the expansion of agriculture and industry, the forest area has shrunk significantly. A portion of the forest, still reminiscent of its past, is now being protected as a national forest.

Chokoria Sundarbans: Once a rich mangrove forest located on the southeastern coast just south of Chittagong, the Chokoria Sundarbans were completely destroyed during the 1980s and early 1990s in order to facilitate shrimp farming. Because of indiscriminate clear-cutting, the area is now beyond restoration.

Despite the gradual deterioration of the country's officially designated and managed forest resources, homestead forestry is now thriving because of a successful people-oriented reforestation campaign. During the 1990s, volunteers planted over 100 million saplings each year. Since then, participation in 'tree plantation fortnight' has virtually become a new social ritual. In addition, the participatory reforestation projects in the coastal *khas* (government-owned) lands as well as along the major roads and embankments have also been quite successful.

Agriculture, Fisheries and Livestock

The alluvial soil and favorable climate conditions make Bangladesh ideal for agriculture, which has historically been the mainstay of its economy. In recent times, commercial agriculture has achieved noteworthy success and the country's grain production for food has doubled since its independence. Early successes in the 1970s and early 1980s can be attributed to the expansion of cultivated areas, often at the cost of wetlands, and the introduction of modern means, such as improved seeds, irrigation, fertilizer and chemicals. At the same time, attempts were made to build up lands vulnerable to floods, thereby increasing the amount of cultivable land. However, in the 1990s two trends were simultaneously observed. The net cultivable area has been gradually decreasing in order to meet the growing demand for land for urbanization and the expansion of the physical infrastructure while at the same time the intensity of cropping has been gradually increasing.

The simultaneous application of these trends has had adverse environmental implications. The increase in cropping intensity without a similar increase in nutrient replenishment has resulted in a gradual depletion of the nutrients and micronutrients found in topsoil. According to many environmentalists, the absence of intercropping and the successive monoculture of high-yielding varieties (HYV) have also played significant roles in hastening this trend. As a result of this nutrient depletion, croplands have yielded less over the past few decades despite the fact that the overall production of grain for food has increased. Moreover, the inappropriate application of fertilizer and other chemicals has accelerated topsoil deterioration. The gradual change in land use, which allows agricultural lands to be used for non-agricultural purposes, is already posing a threat to the food security of the country. Deterioration in the quality of topsoil will only add to the already compounding problems.

The worst long-term environmental impact of modern agriculture in Bangladesh is perhaps caused by very rapid expansion of the use of HYV seeds. The very early success of the 'green revolution' enticed almost every farmer in Bangladesh to raise HYV crops. Very rapidly, almost all local germplasms were withdrawn and it is believed that over 8,000 varieties of grains have been completely eliminated from the traditional crop calendar. It is feared that, given the trends toward monoculture and the socio-economic limitations placed on farmers, an outbreak of infectious disease or other calamity could easily push the entire system toward the threshold of long-term unsustainability.

The implications for allied sectors such as fisheries are also worrisome. The transformation of natural wetlands into paddy fields caused a large-scale loss of open water fisheries. Marine biodiversity experienced a tremendous setback with the introduction of the *bindi* net, with its extremely small mesh size. The catching of juvenile fish and continued fishing during the spawning season have also taken their toll on the open water fisheries. A number of species are believed to have become endangered or even extinct (IUCN, 2000). The introduction of exotic fish species into local waters, some of which prey upon smaller species, has also been detrimental to the available number of local fish varieties.

Despite the fact Bangladesh boasts one of the highest bovine populations in the world, livestock rearing has recently become a very costly affair. Most of the grazing lands have been converted for food production. The traditional livestock feed, straw from rice production, faces uneven competition since it is also being used as a principal source of biomass for rural cooking purposes. Diesel-run mechanical tractors have taken the place of bullocks for plowing. Perhaps because of this, these animals tend to be in poor health and their overall productivity is low. All these factors have discouraged the farmers from investing in livestock as a means to earn a living. But there has been a complete reversal in attitudes towards poultry farming, which is becoming more and more popular in rural Bangladesh, and has achieved sustained growth rates during the 1990s.

Water

To many outside visitors, Bangladesh is seen as a land floating in water. This is, in fact, true during the monsoon months. The constant availability of water and the annual replenishment of topsoil, complete with minerals and loaded with organic material, have allowed agriculture in the region to flourish for many years. On a per capita basis, each citizen enjoys a huge quantum of fresh water, due to the 1360 billion cubic meters of water discharged annually through the GBM system, 93 percent of which passes through Bangladesh. But the perplexing reality is that a large percentage of the land faces moderate to acute moisture stress during the winter months, when there is little or no rainfall. The challenge and the curse of the situation, particularly for maintaining food security for a growing population, is that there is too much water in one season and too little in the other.

Despite having 230 rivers and numerous creeks and rivulets for draining water, the overwhelming amount of water that falls during the monsoons makes it extremely difficult to discharge it all into the sea. Low river gradients and strong backwater effects slow the passage of water through the major channels. It then overflows the riverbank, causing a chronic flood cycle. Such flooding in a delta is, perhaps, essential to help maintain its aquatic-terrestrial exchange and ecological balance. But sometimes synchronization of peak flows in major rivers, accompanied by heavy rainfall throughout the catchment area, causes extreme flooding with far-reaching human and ecological costs. Although embankments were meant to safeguard some areas vulnerable to flooding, they can also cause flooding in the non-embanked communities. It is noteworthy that the worst three of five high-intensity floods in recorded history have actually occurred in the past 15 years (Ahmad, 2000). This finding indicates that the intensity of big floods is on the rise.

During winter there is little to no precipitation, especially in the western part of the country. But the sun alone is enough to trigger high evapotranspiration, which causes acute moisture stress and phenological drought conditions in many parts of the country (Karim, et al., 1990a). Moisture stress in the southern coasts leads to rising surface salinity, and because of this, the potential for the main winter (*rabi*) crops cannot be realized for up to approximately one million hectares of cultivated land. Moisture stress in other parts of the country is tackled by investing in irrigation, and in over 75 percent of the cases, groundwater is extracted for this purpose. Too much of this causes an annual lowering of the amount of piezometric surface in the groundwater tables, which has proven to be detrimental for the expansion of homestead forestry in the northwestern part of the country.

Aside from these problems, water quality remains a key environmental issue in Bangladesh. Prime reasons for water quality degradation include the following: inadequate sanitation and random discharge of fecal matter into water systems, seepage of agro-chemicals into the soil of rural areas; pollution in urban areas due to a lack of adequate, hygienic sewer systems and the expulsion of municipal and industrial wastes; intrusion of salt water into fresh water aquifers; and iron and arsenic contamination of water supplies due to the release of naturally occurring metals into aquifers (World Bank, 1997).

Bangladesh is also facing the serious challenge of supplying potable water to its populace. Contamination of water supplies in both rural and urban areas poses serious problems, and almost all the surface water is currently unfit for human consumption. According to a World Health Organization (WHO) estimate, there were 230,000 deaths directly related to water-borne diseases in 1992 alone. In addition, there were more than 15 million cases of diarrhea, malaria, intestinal worm infestation, and skin diseases caused by water pollution. In the last two decades, more and more people have begun to draw their drinking water from tube wells. Today more than 97 percent of the population acquires its water in this fashion. Unfortunately, since many of these wells are contaminated with high concentrations of arsenic, a large number of people are currently living with the risk of arsenic-related health hazards (Ahmed, 1999).

Energy and Atmosphere

About 65 percent of the total energy requirements of Bangladesh come from traditional biomass energy sources including wood, crop residue and animal waste, with commercial sources supplying the remainder. In 1999, the consumption of different types of commercial energy sources was estimated at 74 kilograms of oil equivalent per year. Natural gas supplied the bulk of Bangladesh's commercial energy, around 70 percent, with a combination of petroleum, hydropower and coal providing the rest. However, the percentage provided by hydropower and coal were quite low. The total number of households that used natural gas and electricity stood at about 3-4 percent and 14-15 percent, respectively. Two to three percent of Bangladesh's households used kerosene for cooking purposes. Due to the scarcity of wood available for use as fuel, a large amount (70-80 percent) of biomass fuel came from agricultural residues and animal dung. All three types of biomass fuels are consumed beyond sustainable limits, and this certainly has adverse effects on the environment. (Islam, 1999).

Recently, there has been much discussion about the potential of the supposedly massive gas reserves in Bangladesh. Debates persist about both the actual size of the reserves and the amount of 'extractable' natural gas they might contain. For the purpose of gas exploration and development the entire nation has been divided into 23 blocs. The government has already leased out all these blocs to foreign companies to invest in the location and recovery of natural gas resources. In 1999, the production capacity of natural gas reached the level of 1000 million cubic feet of natural gas per day. The amounts of this gas that were supplied by the national and international oil companies stood at 76 percent and 24 percent, respectively. The total consumption of natural gas in 1999 was one-quarter trillion cubic feet. Some say that it would neither be possible nor desirable for the total energy requirements of the country to be met by the development of commercial energy resources alone (Islam, 1999). Integrated planning and development of both commercial energy sources and biomass fuels are required for the future sustainable development of energy resources.

The air pollution situation in the country is also quite severe. Air quality in almost all the major urban centers is way beyond WHO standards, especially for the following three pollutants: suspended particulate matter (SPM), sulfur dioxide (SO_2), and airborne lead. SPM levels in Dhaka are 2 to 4 times higher than the national standards, and SO_2 levels there are nearly 5 times higher. It should be noted that the national standards for air pollution in Bangladesh are more lenient than the WHO guidelines. More precisely, SPM and SO_2 levels in Dhaka are 12 and 10 times higher than the WHO standards, respectively, while lead levels are up to 5 times higher than the global standards (World Bank, 1997). The major causes for such high concentrations of pollutants are in fact a combination of factors involving high rates of urbanization, unplanned urban development. Continued construction coupled with the indiscriminate disposal of construction debris, high levels of vehicular pollution, the presence of polluting industries within the city limits, etc.

Two factors are particularly important here. First, there is an increasing trend toward vehicle ownership and use, mainly in order to support the rapidly expanding service sector. Second, industries have been expanding within urban and even residential areas due to the absence of effective zoning restrictions. In the absence of an effective environmentally sound management and inspection system for vehicles, anyone can obtain a fitness certificate even for a vehicle that does not exist. There are also no effective regulations to control industrial emissions.

The exact health costs of air pollution in Bangladesh are yet to be determined. However, they have been estimated for the four largest cities in Bangladesh – Dhaka, Chittagong, Khulna, and Bogra – by using dose-response functions. According to these estimates, nearly 15,000 premature deaths could be avoided if the air quality were on par with the WHO standards. Moreover, rising SPM levels are thought to cause 6.5 million hospital admissions and emergency room visits due to respiratory problems every year. Chronic lung diseases and asthma triggered by higher levels of SPM are estimated to have caused 850 million fewer 'activity days' per year because of respiratory symptoms (World Bank, 1997).

The government has recently taken steps to curb air pollution. Economic incentives are offered to phase out the most polluting two-stroke engines with less polluting four-stroke engines in Bangladesh's rapidly expanding fleet of three wheelers. Moreover, the government imposed a ban on the import of leaded gasoline as a finished product. At the same time, the only refinery in the country that is publicly owned has started to produce lead-free gasoline at a reduced price. Besides, the government has taken initiatives to popularize the use of natural gas in all types of automobiles instead of petroleum products.

Environment and Security Linkages

This section will explore the links between environmental challenges and security concerns in Bangladesh. In order to organize this discussion around a conceptual template, we will discuss the links between environmental and security concerns at four distinct levels:

- Threats at the national level, based on sovereignty concerns.
- Threats to human security, based on the degradation of natural resources upon which livelihoods depend.
- Threats to human health.
- Threats to ecological security.

Environmental Challenges and National Security

Sometimes an action taken either by the state or by a group of individuals can threaten social harmony and peace, or even spur a situation which could potentially threaten the sovereignty of the state itself. Policies made by the government can sometimes cause the destruction of an area's environmental harmony, thereby threatening state security and sovereignty. This is an expensive lesson that the people of Bangladesh have learned over the past thirty-five years; it has incurred huge social, economic and environmental costs.

In the early 1960s it was decided to build a dam on the Karnaphuli River, located in the hilly south-eastern part of the country, which would produce hydroelectricity. The government of Pakistan took the initiative and implemented the plan despite the protests and resistance of the local people. The dam inundated a total of 68,800 hectares of land, including the ancestral homelands of a number of indigenous tribes. Even the palace of the tribal king and all the religious holy places were inundated, along with the shanty houses and farming lands in the valleys. The ecosystem of the area of intervention changed overnight. The production system for the survival of the indigenous people was disrupted and people lost their livelihoods.

Following independence, a decision made by the new Bangladeshi government virtually denied the indigenous people the right to own their own land, since they had never registered it under the legal framework of the state. This decision triggered a large-scale armed conflict between governmental forces and the indigenous people. The unbalanced war forced the tribal people to take refuge in deep forests, sometimes across the border in neighboring India. Meanwhile, the armed forces denuded the pristine forest as a means of uprooting the 'rebels' altogether. The conflict resulted in the complete destruction of the ecosystem, which had been the basis of the survival of the indigenous community. Eventually they threatened to 'free' their motherland and wanted to gain independence by dividing Bangladesh. The conflict continued for about a quarter of a century and threatened the very sovereignty of the nation. This example demonstrates how policies adopted by the state can cause the wide-scale destruction of the environment and potentially threaten the sovereignty of a country.

Since the mid-1980s, investors have started to take control of the coastal floodplains, converting the croplands into shrimp-beds. Such a drastic change in land use, which required deliberate salinization, has been at least ten times more profitable in financial terms than it had previously been. Though traditionally shrimp has been cultivated in harmony with the ecosystem of the area, the unplanned nature of renewed shrimp cultivation resulted in excessive salinization due to prolonged inundation of lands with saline waters, which is injurious for crop production. A quick investment return lured rich investors from afar to forcefully flush the neighboring land with salt water, making it less useful for traditional agriculture, and finally taking it under their command by driving the poor farmers away from their homelands. This change from traditional agriculture to shrimp cultivation has also resulted in a loss of employment for the local people because shrimp farming is comparatively less labor-intensive. While investors have enjoyed good harvests, the livelihoods of poor farming families have been seriously threatened. Such a change in land use has increased social inequality and escalated real conflicts between poor farmers and the musclemen employed by the investors. These conflicts have disrupted local peace and social harmony, and sometimes even required state intervention in order to maintain the peace.

Environmental Challenges and Livelihood Security

Historically speaking, the population of what is now Bangladesh enjoyed relatively good living conditions until the middle of last century. Until then, the natural resources were quite sufficient to sustain the population. Even foreign visitors found it to be a prosperous place. Unfortunately, the subsequent population boom stretched the limits of the natural resources and it became increasingly difficult for the population to maintain their livelihoods, which often depended on these resources. Except water in monsoon season and newly-discovered gas deposits, its per capita endowment of productive natural resources is today one of the lowest in the world. This has led to the overexploitation of some natural resources and put pressure on the resource base, which maintains the livelihoods of over 70 percent of the population of Bangladesh, whether directly or indirectly.

Land is perhaps the most precious resource of all. Between 1971 and 2000, the per capita land availability declined from 0.20 hectares to 0.11 hectares. Due to this acute shortage in land availability, the nation's food security is threatened. To expand land coverage for agriculture, the wetlands were deliberately drained. This change in land use produced good results in terms of increasing production, but it has severely restricted the amount and efficiency of nurseries for fish and other aquatic species. Both open water fisheries and fishing communities have suffered because of it. In particular, the lost opportunities due to an inability to fish in shallow bodies of water in the dry winter have caused much hardship and insecurity for these fishing communities.

Simultaneously, some privileged areas were made virtually flood-free by the construction of embankments. The fishermen not only lost the opportunity to catch fish from areas which had previously been flooded, but they also could not gain an advantage from investment-intensive capture fisheries located inside the embankments. A large number of anglers were forced to change their traditional profession due to acute livelihood insecurity. On the other hand, the loss of fish habitats has resulted in a decline in marine biodiversity, which would certainly affect the long-term sustainability of open water fisheries and the people who depend on them.

People inside the embanked areas took full advantage of flood-free agriculture, since they could now grow crops both in winter and in the monsoon season. The development of flood-tolerant plant varieties also took place in flood-prone areas. While this has indeed yielded economic benefits, it has also meant an increase in agricultural intensity throughout the country. The time between two successive growing seasons has become very short and is often not long enough for major nutrients and micronutrients to be replenished in the topsoil. As a result, the overall concentration of nutrients and micronutrients in most of the nation's croplands is on the decline.

With the advent of modern advances such as high-yielding seeds, irrigation, fertilizers, insecticides and herbicides, among others, farmers quickly adapted to new practices and reaped initial benefits. From the early 1970s, food grain production rose steadily, and food grain security was maintained. Since rice, the national staple food, was grown more than other crops, farmers tended to grow more paddies, and the land allocated for other crops has steadily declined. However, this over-emphasis on food grains has caused a serious imbalance in many people's daily diet, especially those of the poor farmers, leading to health concerns. The rate of malnutrition among children below five years of age is even higher than in the drought-prone Sub-Saharan Africa, and is thus posing a serious threat to the formation of human capital the in future.

Over 60 percent of the total energy demand in Bangladesh is currently being met by biomass sources, mainly agricultural residues and wood. Since independence, there has been no significant advancement in terms of providing energy to the rural poor. They are deprived of commercial energy services and they have no other choice but to meet their needs using biomass sources. Their energy security has been stretched so far that they burn dried cow dung, which had previously been applied to croplands as fertilizer. Moreover, they deliberately take all the dry biomass from the fields after the harvest, including the stubble, and do not replenish nutrients to the soil by mulching those materials during field preparation. As a result, the concentration of available organic matter in most of the croplands has gone down significantly. The link between the lack of biomass fuel and the decision not to replenish the soil has been identified as a threat to the long-term sustainability of agriculture in Bangladesh, and therefore to the food and livelihood security of the Bangladeshi people (Ahmed, 2000).

Over the past three decades the actual tree cover in government-owned forests has declined significantly. During the 1980s, the rate of deforestation was approximately 37,600 hectares per annum (MEF, 1995). Much of this might have occurred in order to meet the growing demand for wood for fuel in rural areas. Although Bangladesh is endowed with a healthy reserve of natural gas, the effective energy security of a large number of rural people will depend solely on traditional energy sources. Environmental constraints can only aggravate the scarcity of biomass energy, threatening the livelihoods of the rural poor.

The worst form of livelihood insecurity is caused by natural disasters. The country is tragically and repeatedly at the mercy of multiple natural calamities. Floods, riverbank erosion, cyclonic storm surges, droughts, and salinity intrusion are the most common phenomena that attack Bangladesh. Natural disasters destroy the nation's economy, infrastructure, and most importantly, its potential for development. These hazardous events wreck havoc on the livelihood of the poor, destroy standing crops, take away their assets and belongings, and make living itself a disaster (Haider, et al., 1992). People find no time to resettle as the next natural disaster hits them even before they can recuperate from the losses of the first.

All these forms of environmental destruction put daunting pressure on the livelihood security of millions of Bangladeshi people. It is mostly because of a combination of these factors that many people are relocating *en masse* to urban areas and are often forced to accept inhuman living conditions there.

Environmental Challenges and Human Health Security

A degraded environment causes significant threats to human health security. The gravest concern in this regard is arsenic contamination in groundwater systems. Ninety-five percent of the population of Bangladesh get their potable water from groundwater sources. As previously discussed, the amount of arsenic contamination in this groundwater is quite alarming. Even though the tube wells that showed the presence of arsenic in water above the national limit are now marked, people in the area have no alternative but to continue to drink from them. Because of their continued exposure to arsenic, incidences of arsenic poisoning and related health hazards are on the rise (Ahmed, 1999).

The very reason that the government wanted to provide drinking water from tube wells is that the surface water bodies have been heavily contaminated with pathogens. In the past, people drank the surface water without receiving treatment and suffered from various water-borne diseases. The major problem was that due to improper sanitation facilities, people used to defecate out in the open. In monsoon season the surface water systems easily became contaminated with fecal coliform and other pathogens. Therefore, drinking surface water caused outbreaks of diarrhea.

The temperate climate, excessive humidity and frequent extreme weather events make Bangladesh extremely vulnerable to tropical diseases. Vectors are found in abundance. Interplay of all these factors along with poor nutritional intake is responsible for these large-scale health problems. It is no wonder that in terms of infant mortality and life expectancy, Bangladesh lags behind many other developing countries (UNDP, 2000). The situation is dire enough to warrant being termed a threat to the human security of the Bangladeshi population at the most intimate level.

Environmental Challenges and Ecosystem Security

A less obvious, but equally pressing, challenge for the nation is the looming threat to the continued security and sustenance of the vital ecosystem on which the environment and the people of Bangladesh depend. Let us consider one key example of how ecosystem changes are linked to the problematique of environment and security.

In 1975, a barrage was commissioned by India on a trans-boundary river, the Ganges. The location was 18 kilometers upstream of the border between Bangladesh and India, along the Rajshahi district. The Farakka barrage was aimed at restoring the navigability of the Bhagirathi-Hooghly river by diverting 1,133 cubic meters of water per second (m^3/s) from the Ganges. The diversion of water through the barrage changed the hydrology of the entire Ganges-dependent area in Bangladesh, especially during dry periods. The average discharge of the Ganges in April (the lean season) in the Hardinge Bridge area decreased from 1,987 m^3/s in the pre-Farakka period to 1,044 m^3/s in the post-Farakka period (Mirza, 1998).

Due to a gradual decrease in low flow in the Ganges, some of its tributaries virtually died, and due to changes in sedimentation patterns, others (such as the Gorai) choked. As a result, salinity penetrated inland through surface water systems (see Table 4.1). The agro-ecological characteristics of nine districts have been completely changed over the past quarter of a century because of this project. Open water fish production has declined in the entire southwestern region due to shrinkage in surface water systems. Because of the intrusion of salinity, over 800,000 hectares of potential dry season paddy lands had to be forfeited each year (Karim, et al., 1990).

Table 4.1

Pre- and post-Farakka barrage average monthly maximum salinity at four stations in southwest Bangladesh

Salinity at...	**March**		**April**		**May**	
	Pre-Farakka	Post-Farakka	Pre-Farakka	Post-Farakka	Pre-Farakka	Post-Farakka
	DS/m (at 25°C)					
Khulna	0.47	8.31	1.63	12.15	1.51	11.21
Goalpara	0.75	4.42	1.32	7.42	0.79	5.46
Chalna	8.95	17.31	8.68	21.93	12.00	19.01
Mongla	7.50	11.08	11.80	17.15	13.50	17.10

Source: Karim et al. (1990)

The worst impact of all has been on the Sundarbans, which is the largest patch of productive mangrove forest in the world. Due to a lack of flushing through the Ganges and Gorai rivers and high salinity contents, the regeneration process of the forest was disturbed. The southwestern corner of the forest bordering India lost the high-value timber producing species *sundri*, which was replaced by low-value shrubs. Potkin (1987) reported that lack of adequate flows of fresh water to the Sundarbans caused the degeneration of freshwater plants. The forest's ecosystem, which used to be rich in biodiversity, suffered greatly because of the upstream diversion of the Ganges. In a 1982 study, the Bangladesh Department of Forests claims that a staggering 1.43 million cubic meters of *sundri* timber was lost between 1976 and 1982. A very conservative estimate of the timber loss as a result of the Farakka dam project is 320 million US dollars (Swain, 1996). Moreover, the forest was the means of livelihood security for about 3 million people, whose loss cannot be counted in dollars.

This also links to our earlier discussion of how the perennial wetlands of Bangladesh have now been encroached upon by the expansion of crop agriculture. These were highly productive wetlands with a high density of aquatic biodiversity. Many of these aquatic species have lost their only habitat because the wetlands dry up in winter. For example, the *haor* ecosystem in the northeast was a set of water bodies which depended on the year-round presence of water. Now this habitat, which is vital for some important aquatic species, is fragmented and a number of these species are on the verge of extinction (IUCN, 2000). Habitat fragmentation is a direct consequence of environmental degradation in northeastern Bangladesh.

The most important fish species for Bangladesh is Ilish (*Hilsa Ilisha*), which was previously found in abundance in the Meghna and Ganges rivers. Now the annual production of this particular species has declined drastically and the fish is no longer available at costs that can be afforded by poor or even lower middle-class families. Part of this low production level may be attributed to catching the juveniles, called *jatka*. But many believe that it is due instead to deteriorating water quality as well as increased sedimentation in the rivers. Increased sedimentation blocked the regular communication channel of Ilish fish from the Bay of Bengal to these rivers. Although the level of industrialization is still low, these industries do not follow the laws regarding effluent discharge; they pollute water bodies indiscriminately. In addition, the inappropriate application of chemicals in agricultural fields makes sure it stays on the ground to be transported into surface water systems in monsoon season.

In short, when environmental deterioration leads to ecosystem destruction, the impacts are not only compounded but are often part of an enduring downward spiral that invariably leaves communities and people more impoverished and less secure.

Future Environmental Threats and Insecurity

In this final section, we turn our attention from the significant questions of today to the looming challenges of tomorrow. What are the issues that are likely to cause wide-scale environmental degradation and therefore pose threats to the security of Bangladesh and its people in the future? This section identifies four key issues: urbanization and industrialization, pollution in trans-boundary water systems, toxic waste, and climate change.

Rapid Urbanization and Industrialization

Long-term projections suggest that Bangladesh will be rapidly urbanized (BCAS, 1998) and that the urban population will surpass the rural population by the year 2020. Practically, the country does not have the financial and/or physical capacity to deal with, or even plan for, rapid urbanization. As a result, urban management will remain a problem and it will create huge environmental issues. Losses in production are likely to increase if the current rate of urbanization continues. In addition, if industrial activities are increased in the near future – as they most probably will be – and if the effluents are not discharged after proper treatment, it is very likely that the effects of pollution on wetlands and surface water will be quite detrimental. An example of the kinds of problems that may be in store for Bangladesh's people is the increased salinity in the Bhairab River in southwest Bangladesh that forced the management of a paper factory to collect freshwater from 100 kilometers north. The financial and ecological costs of such activities are quite extreme and questions of water quality could in this way threaten the very future of industrialization in Bangladesh.

Pollution in the Transboundary River Systems

Bangladesh shares 57 transboundary rivers including 54 with India and three with Myanmar. In all cases except one, Bangladesh is the downstream country on the river. Therefore, in essence Bangladesh is dependent upon its neighbors, especially India, for the water that runs in its rivers. Since more and more urbanization is likely to take place along these rivers, it is also likely that water quality will further deteriorate. Under such circumstances, all the production systems involving water will be at risk of quality failure and will face the challenges that were discussed earlier. The fact that these rivers are transboundary water bodies makes them especially susceptible to national security concerns. In many cases, the forces of urbanization and industrialization are likely to be as strong upstream (i.e., before the rivers enter Bangladesh) as they will be in Bangladesh itself. This could raise a number of bilateral international security concerns relating to the quantity and quality of water that would enter Bangladesh. Given the contentious history of such transboundary water disputes in the region and elsewhere, one can be relatively sure that these transboundary river systems are likely to be amongst the areas where environmental and security concerns will be most prominent over the next two decades and beyond.

Toxic Waste Dumping

The dumping of toxic waste is a growing international concern. In recent years there have been several attempts to illegally import and dump toxic wastes within the borders of Bangladesh. The international trade in toxic waste, although illegal, is often carried out under the cover of legal provisions and tends to work best where the legal framework is either incomplete or is weak in implementation. Since industrialized and rapidly industrializing countries are generating vast and increasing quantities of toxic waste, and since the cost of disposing it in the countries where they are made is too high, the incentive remains to dump them in poorer countries. It is expected that the multinational companies will tend to dump in places where there is poor implementation of legal provisions. Bangladesh will therefore be under the constant threat of toxic waste dumping. The implications of such a likelihood, either as illegal dumping or legal 'waste trade', are severe not only because of the environmental impacts but also for security reasons, including concerns about security at the state as well as the human levels.

Climate Change

The most serious future threats to Bangladesh are likely to come from exogenous sources. It is now accepted internationally that Bangladesh will be highly susceptible to the impacts of global climate change and rising sea levels. A combination of all these changes is likely to increase both the number and the intensity of disasters, as well as to further reduce Bangladesh's agricultural potential, challenge the offshore gas mining activities, and make millions of its already poor people destitute (Huq, et al., 1998). Bangladesh's contribution to global climate change is negligible but it will face a disproportionate amount of adverse multi-dimensional impacts. Climate change is likely to aggravate the problems of environmental degradation and perhaps even serve to destabilize all aspects of security. It is anticipated that the vulnerability of the country to climate change might be so high that a large number of people will seek to migrate elsewhere, either inside the South Asian region or beyond. This will, in turn, serve to destabilize the social and political peace in this region (Ahmad and Ahmed, 2000).

Conclusion

Many developing countries today face the dilemma of growing expenditures on military hardware accompanied by steadily decreasing national human security. They have adopted security strategies based on military considerations while ignoring fundamental socio-economic needs. Today, the concepts of human and environmental security call for a move from *technical* and *military* solution of security to *socio-economic* and *environmental* considerations. Bangladesh, with a burgeoning poor population, a massive number of children under malnutrition, and with a rapidly depleting natural resource base has an uphill task before it and not an impressive record behind it. A radical change in political thinking will be required along with comprehensive, coordinated, and cohesive measures to deal with the problems. Most importantly, countries will need to start defining national security in terms of the human security of their inhabitants.

Chapter Five

Exploring the Environment & Security Nexus in Sri Lanka

SARATH W. KOTAGAMA

DISCUSSIONS ON THE environment have come a long way since the the early 1960s. Environmental concerns have penetrated almost every area of decision-making and have become a major cause of concern for society. This infusion into different academic disciplines and sectors of governance has invariably resulted in environmental concerns being differently interpreted with respect to each academic discipline or policy sector.

The use of nature and the environment for supporting the well-being of humans has become a major component of what is widely recognized as 'development'. However, even where there exists the realization that any use of natural resources and the environment must be based on basic principles of ecology and environmental science, it is also quite clear that the multiple and complex dynamics of economics, politics and social structures also influence this use. In combination, these factors lead to a complicated entanglement resulting in the environment as an important aspect of the new human security problematique. The complexity, enormity and import of this problem are particularly pressing in developing countries such as Sri Lanka.

While the economic, political and social pressures that shape the discussions of security in general are themselves complicated and less than well understood, the links between environment and security become particularly muddled and difficult to understand because of new layers of complexity added by our lack of full appreciation for environmental dynamics. Despite recent advances in knowledge, there persists a significant lack of clarity in our understanding of ecology and environmental science; of its applications in resource use, protection and conservation; development and sustainability; etc. The objectives of environmentalism – emanating from advocacy attempts aimed at environmental protection, conservation and sustainable use – have themselves lost direction and purpose as a result of this confusion. It has also led to a state of tension within and among disciplines, resulting in the development of a new concern – 'environmental security'.

To explore the implications of these new concerns for environmental security in Sri Lanka, we begin by briefly exploring the environmental history of the country leading into a quick review of the current state of Sri Lanka's environment. This will be followed by a set of case studies of recent conflicts and controversies that highlight the various facets of environmental security (and insecurity) in contemporary Sri Lanka.

Sri Lanka: Historical Setting

Sri Lanka's proud and illustrious history spans more than 2,500 years. Sri Lanka is an island in the Indian Ocean with a total area of 65,610 square kilometers. Pear-shaped, its central area is comprised of hills rising up to heights of 3000 meters. This central area comprises about one- third of the total land area. The rest is rolling terrain interspersed with mountain islands and rock outcroppings. Radiating out from the hilly central area are over one hundred river systems.

The island lies south of India, on the same continental shelf. The separation between the countries at its narrowest point is about 40 kilometers. This proximity to India has, of course, had an impact on Sri Lanka. However, the insularity afforded to it as an island has permitted a rich and unique culture to develop. The most pronounced outside influence has been Buddhism, which came into the country around the third century B.C. Sri Lanka's placement at the tip of the Indian peninsula has made the island an important point in ocean trade pathways. As a major port of call in transoceanic trade over the centuries, Sri Lanka was able to come into early and repeated contact with major civilizations over the ages including the Greek, Roman, Arabian, Chinese, Islamic and European powers. These cultural links have left their impact over time.

Based on some of the traditional historic chronicles (such as the *Mahavamsa*, or 'Great Chronicle') it can be seen that Sri Lanka's civilizations flourished in the dry zone from the fourth century B.C. to approximately the twelfth century. These documents reveal that the nation's population may have been in the region of 20 million (slightly more than today's, which is nearing 19 million) at a time when Sri Lanka was at its developmental peak: 8-12 A.D. The background to this success is attributed to hydraulic technology, based on the invention and adoption of some of the best water management practices of the time. This technology would probably be described today as sustainable development technology (Fernando and Denis, 1980).

The collapse of this heritage has been attributed primarily to foreign invasions that led to a series of interconnected events. The technology and management systems that had evolved to maintain and nurture the natural wealth of the island were destroyed. As a result, insecurity and mismanagement set in, leading to chronic neglect of the essential systems. Among the many negative impacts of these events was the increased prevalence of vector-borne diseases such as malaria (NARESA, 1991). The compounded effect of all of these was the erosion of a once glorious civilization. Here lies a lesson from the past on ways in which human security can impact environmental conditions and the very survival of civilizations.

Political disintegration commenced around the thirteenth century, and by the sixteenth century settlements had been established in the Wet Zone and Hill Country. Since then, invasions from Europe -- the Portuguese, the Dutch and finally the British – left much of the land desolate and abandoned. After 1815, the British commenced a systematic process of land use change in the country. These changes were primarily centered in the wet zone: coffee plantations, followed by rubber and then tea. It was only in the late 1930s that the dry zone areas were re-colonized. These land-use changes have revolutionized land practice in the country and set up regimes of settlements in both the wet and the dry zones. This has had a tremendous impact on the current environmental profile and also on the options that are (or are not) available to Sri Lanka for future development.

Land settlements, initiated by constitutional changes, began in the dry zone in the 1930s. According to the Land Commission of 1927-29, this process of parceling out land to the people in the dry zone was termed 'colonization' with the objective of the 'preservation of the peasantry' (Peiris, 1996). This opened possibilities for the growing population to avail of opportunities for economic advancement and social well-being. As part of this process, many of the abandoned reservoirs were rehabilitated. The first settlements were built in 1936, and by 1948 eleven major schemes were completed, with about 16,000 families settled. This number grew to nearly 86 more schemes and 49,000 families by 1978. Most of these settlements were in forested areas, and even today there exist environmental concerns in these areas that impinge on human security. One concern that has remained relevant is the case of the so-called 'Human-Elephant Conflict', which will be discussed later.

Although population growth has slowed down recently, it was fairly rapid during the second half of the twentieth century, increasing from 6.9 million people in 1948 to 17 million in 1990. It is estimated to be around 19 million today. Early development emphasis in this post-independence period was aimed at increasing the amount of available land for agriculture. After 1970, however, the trend toward industrialization became more pronounced, gaining further prominence after 1977. These development activities have created centers of high population density, often accompanied by concerns about human security and environmental quality. Some of these concerns can also be classified as environmental security issues.

At the same time, international pressures for greater environmental awareness began to have an impact on Sri Lanka. The 1960s saw increasing concerns in the developed world about the links between development and the environment. Activism on this issue led to the initiation of Earth Day in the United States in the 1970s, and global policy soon followed with the United Nations Conference on the Human Environment (UNCHE) held in Stockholm in 1972. In response to the confluence of growing domestic interest and these international events, the first Sri Lankan environmental NGO (non-governmental organization), the Environment Society of Sri Lanka, was established in 1970. With a very restricted membership, this organization worked alongside other nature-based NGOs to lobby for the establishment of a governmental authority for the protection of the environment. These actions, with assistance from the United Nations (UN), bore fruit in 1980 with the establishment of the Central Environmental Authority under the Prime Minister, through the National Environmental Act of 1980. With subsequent amendments, the National Environmental Act remains the major piece of legislation regulating the environmental activities of Sri Lanka. By 1989, a good three years before the 1992 Rio Earth Summit, Sri Lanka had created within its government a separate Ministry of the Environment.

Sri Lanka: State of the Environment

Having established the historical context, we will now review the current state of the environment according to selected areas of concern as reflected in the National Environment Action Plan (NEAP) of 1998.

Water, Land and Air Pollution

Water has been the major resource base for Sri Lanka's ancient and modern civilizations. The traditional technologies developed around the use of water for irrigation are unparalleled, and the *hydraulic civilization* of the ancient kings is still the backbone of the present agricultural system. Indeed, one could argue that in some respects, Sri Lanka is still a hydraulic civilization. At present, diverting and/or augmenting water to basins irrigates more than 650,000 hectares of paddy land. Eighty-five percent of the nation's electricity needs are currently met by hydropower. In addition, groundwater reservoirs are used heavily by households, factories and farms.

The heavy extraction of water, the increasing use of chemicals in agriculture, rising sediment loads, and increases in industrial and household wastes have all meant higher stress levels and pollution in Sri Lanka's water bodies. More than 30,000 man-made bodies of water lie in various stages of disrepair; many are overgrown with aquatic plants, which leads to eutrophication. The main water reservoirs in a number of key urban areas, such as the Beria Lake in Colombo, the Lunawa Lake in Moratuwa, and Lake Gregory in Nuwera Eliya are becoming a risk to human health due to their high pollution levels.

The ground water has also become heavily contaminated with agro-chemicals and in some cases its use has had to be restricted by governmental regulations. Sri Lanka has nearly 300 high-polluting industries that tap the water sources for their industrial needs. These industries also routinely expel their waste into water systems. These environmental pressures have led to problems with both the quantity and quality of available drinking water. A manifestation of the dwindling quality of water in taps and wells is the emergence of a thriving bottled water industry. While water quality directly impacts the health and security of the population, it also has indirect economic implications on human security because it raises the actual cost of obtaining drinkable water. Not surprisingly, the biggest burden falls on the poorer segments of society.

With increasing urbanization and industrialization come increasing concerns for air and water pollution. During the last decade, industrial pollution has been particularly problematic. Although largely confined to industrial estates at the moment, the amount of emissions and effluent being produced and the trends of future growth are quite worrisome. Principal areas of concern remain in and around Colombo and other major urban centers. Another key contributory factor to air pollution is vehicle emissions. A vehicle fleet growing at 7 percent annually and using leaded petrol increases the stresses. Even though the conditions are not as bad as in many other Asian capitals, some areas within the city of Colombo exceed World Health Organization (WHO) standards with respect to particulate matter (NEAP 1998).

Table 5.1
Catchment soil erosion rates, Sri Lanka

Catchment	Area (sq. kms)	Period	Sediment Yield (mt/ha/year)
Upper Mahaweli (above Peradeniya)	1,167	1992/93	5.8
Upper Mahaweli (above Polgolla)	1,292	1982/92	8.9
Nilambe Oya	60.5	1991/92	0.6
Victoria	1,891	1985/92	3.4
Maha Oya	476	1985/92	9.4
Uma Oya	94	1993/94	10.6

Source: Wallingford (1995)

Like water and air, land is also being polluted and degraded at significant rates. Land degradation resulting from improper and often intensive land use practices has resulted in a serious problem of soil erosion (Wallingford, 1995). This, along with high levels of fertilizer use, has accelerated the degradation process. The soil condition in the mountain wet zone is considered quite acute. The production potential of the arable soil in the rest of the country is also generally degraded resulting from the extensive use of agro-chemicals. In many areas, these practices have already destroyed the soil micro- and macro-organism base for healthy soil generation. Examples of the extent of the problem include the facts that in the Upper Mahaweli catchment the extent of erosion is to the tune of 132,000-833,000 tons per year, the sediment load of the Mahaweli Ganga is around 15 million tons during the period 1952-1982, and the erosion in the Maha Oya catchment has been between 400-913 tons per hectare per year. The soil erosion rates in the various catchments are of particular concern (see Table 5.1).

Solid and Hazardous Waste and Environmental Health

With increases in urbanization and living standards, the increase in solid waste has become another problem that requires urgent attention. It is estimated that the city of Colombo alone produces nearly 700 tons of solid waste every day. Table 5.2 provides the waste generation figures for the leading urban centers in the country. The western province, which has the highest population density, has approximately 58 waste dumps, all of which are classed as open dumps with little or no sanitary regulations. Uncontrolled burning, leaching and putrification of wastes at these dumps lead to many health hazards as well as to long-lasting environmental contamination. With an increasing demand for land and the accompanying increase in waste production, Sri Lanka is faced with a real waste crisis which is only likely to become more severe with time. Within this waste stream, the growing component of hazardous waste poses a particular challenge. According to some estimates as much as 45,000 tons of hazardous waste are produced in Sri Lanka every year, mostly by hospitals. In the absence of any meaningful waste management regime, nearly all of this is dumped openly and indiscriminately, thereby endangering both human life and environmental integrity.

In the area of health, Sri Lanka, like most other developing countries, has invested more attention in curative interventions rather than preventative interventions. Although this has yielded results in that the average life expectancy in the country has now reached 71 years, the relative lack of focus on preventative interventions has begun to create problems. In particular, the occupational health situation in relation to agriculture, industry and other occupations remains a serious environmental health threat. Pesticide poisoning among the young in rural areas remains high. Health problems arising from less than satisfactory water and sanitary conditions are acute in urban low-income settlements. In addition, vector-borne diseases remain a persistent environmental health concern (NEAP 1998).

Table 5.2
Waste production in major cities of Sri Lanka

Local Authority	Population	Area (sq. kms)	Gross Weight collected per day (in kgs)
Colombo	800,000	37.00	680,000
Dehiwela/Mt.Lavinia Municipal Council	247,000	21.60	150,000
Galle Municipal Council	143,000	6.5	45,000
Matara Municipal Council	46,000	8.1	23,000
Kandy Municipal Council	145,000	29	102,000

Source: Ministry of Forestry and Environment (1999)

Loss of Biodiversity

Historic and recent events have had a profound effect on the flora and fauna of the country. Amidst the large-scale changes in the socio-economic and political structures, the natural forest cover has been reduced to around 23 percent of an area that was 80 percent in 1881 (Ministry of Forestry and Environment, 1995). Not inconsequentially, the population of the island has grown in this period from 2.7 million to 18.3 million. Many of the other natural systems have also degraded. Table 5.3 provides a snapshot of the remaining natural forests, by type.

Sri Lanka still boasts of being among the top countries in Asia for biodiversity; it is one of the world's 17 recognized mega-diversity centers (Myers 1991). As detailed in Table 5.4, 13 percent of its land area has protected status (NEAP, 1998; SOE, 2002). In addition to this rich biodiversity base, Sri Lanka has inherited a rich culture that gives greater value to the nurture of its biodiversity. Nevertheless, there exists significant concern about the rate at which Sri Lanka's biodiversity is either being depleted or degraded due to neglect. On the one hand there is a sense that the support received for biodiversity conservation is less than desirable; on the other hand, concern is raised about the perceived conflicts between biodiversity conservation and human development. These conflicts lie at the very heart of the environment and security nexus in Sri Lanka. A case in point is that of conflicts between humans and elephants. During the 1990s, the number of elephants killed by humans as well as the number of humans killed by elephants were both increasing.

Table 5.3
Natural forests by type, Sri Lanka, 1992

Forest Type	**Total Forest Area (hectares)**	**Percentage of Total Land Area**
Montane	3,108	0.05
Sub-montane	68,616	1.04
Lowland wet evergreen	141,506	2.14
Moist evergreen	243886	3.69
Dry mixed evergreen	1,090,981	16.49
Riverine	22,435	0.34
Mangrove	8,688	0.13
Total extent of closed canopy forests	1,579,220	23.88
Open-canopy/ Sparse forests	464,076	7.01
Total Forest Cover	**2,043,296**	**30.89**

Source: Ministry of Forestry and Environment (1995)

Unsustainable Use of Natural Resources

As with biodiversity, Sri Lanka is rich in natural resources such as forests, fisheries and minerals, but there is growing concern that continued unsustainable use of these resources could harm both the human security of those whose livelihoods depend on these resources and the ecological integrity of the island's resource base.

Table 5.4
Protected Areas by Percentage, Sri Lanka, 1992

Land Designation	Total Area (hectares)	Percentage of National Land Area
Protected Land Area	966,225	14.00
Area Administrated by Department of Wildlife	811,511	12.30
Area Administered by Department of Forest Conservation	154,947	1.70

Source: Ministry of Forestry and Environment (1995)

The forestry sector has, with the second Forestry Sector Master Plan of July 1995, taken a positive step toward integrated management with the involvement of the private and public sectors as well as greater emphasis on participatory and community-based decision-making with multiple stakeholders. However, given the hundred-year history of top-down, command-and-control management of the forestry sector in Sri Lanka, operationalizing these new approaches in practice will be a formidable challenge. Despite this good beginning, conflicts over boundaries and resource allocations between different stakeholders still persist. It is these conflicts, lying at the nexus of environmental and human security concerns, which will be the ultimate test of the new master plan.

The situation regarding minerals is even more precarious. Currently, minerals are exploited with little concern for environmental impact. Institutions for managing the use of this resource have yet to be developed and the legislative process remains nascent. The Eppawala phosphate mining controversy, which we will discuss later, highlights the potential for friction along the environment and security nexus developing in this area.

Fishery, which is a mainstay not only of the economy but also of the Sri Lankan lifestyle in many regions, is an equally fertile ground for environmental security concerns. With increasing populations and growing demand, this resource has been under increasing pressure. For example, the total fish production for the country rose from 235,750 million tons in 1985 to 260,100 million tons in 1990 in what has been a steady but unsustainable growth in fish consumption (NARESA, 1991). Near-shore fishing has been further affected by land-based pollution spilling into the spawning ground, which forces fishermen to place increased stress on off-shore resources. This, of course, is an obvious recipe for unsustainability. On the one hand, the new focus on deep-sea fisheries does no more than delay the problem, rather than solve it.

Additionally, deep-sea fishing is more capital intensive and serves to exclude the poorest fishermen, endangering their human security. The changing nature of the industry, particularly its now more capital intensive nature, is driving the average fish prices higher and is therefore also affecting consumption patterns. One may add here the continuous extraction of ornamental fish – marine and fresh water – which has been of real concern to many ecologists for the pressure it imposes on biodiversity. The resource has been severely affected, especially in the fresh water sector. Even with an average mortality rate of 35 percent per shipment, the practice continues. The loss of many fresh water sources can be directly attributed to increased pressures of exploitation and is further aggravated by large amounts of agro-chemicals leaking into the water sources.

Future Trends

Sri Lanka has had an average economic growth rate of 3-4 percent per year during the last decade and a half; it is expected to maintain at least this rate of growth over the next decade. In addition, the population has been growing at around 1.2-1.3 percent. It is expected to continue growing at this rate, although the country's population is expected to become increasingly aged over time. The population concentration continues to increase in the wet zone in the western region. As a younger generation takes to urban and industrial employment, the structure of society is making a shift away from agriculturally-related employment. This shift will inevitably change the face of Sri Lankan agriculture and its productivity. The pressures of globalization notwithstanding, significant changes are already taking place in Sri Lanka's social, economic and environmental structures which are likely to have significant ramifications for the environmental and security nexus in the country. Some key trends are listed here as a sampling of the forces at work (NEAP, 1998):

- Sri Lanka's population remains unevenly distributed with over 75 percent still in the rural areas.
- The balance is in rapid flux and the population of urban centers is expected to rise to 40 percent of the total by 2020.
- This shift will bring with it an attendant shift in consumption patterns; the per capita consumption of natural resources is expected to rise with increasing income levels.
- Energy demand, in particular, is expected to rise significantly -- at rates of up to 10 percent per annum by some estimates. The nation's energy supply is currently mostly clean due to its heavy reliance on hydroelectric power. However, increasing demand will most likely imply a shift towards thermal energy and more heavily-polluting energy sources.
- An increasingly urbanized population will face a new set of issues that urban municipalities are currently not ready to deal with. For example, it is expected that not only will the amount of solid waste generated from households increase, but it will contain increasing amounts of non-biodegradable components.
- Vehicle traffic and its environmental and health impacts are likely to continue growing at 7 percent or more, per annum.
- The per capita income is projected to reach one thousand US dollars soon, but the distribution remains highly uneven.
- Poverty will continue to be a prevalent factor, limiting both developmental and environmental options available to Sri Lanka and its people.

These and other similar issues will shape the future direction of Sri Lanka's development and of its people's human security. A very significant contributory factor to the security issue will be the level and shape of institutional devolution of governance that has currently been negotiated in all sectors of society. Pending constitutional and legislative changes could alter the very nature of the manner in which decisions regarding ecology and human security are made in Sri Lanka, as well as the methods by which conflicts concerning these decisions are resolved. Institutional changes may, therefore, have as much influence on the future of the environment and security nexus in Sri Lanka as any aspect of the physical environment. Issues of security are brought into sharper relief given the ongoing ethnic conflicts in the north and east, influencing not only the security of the individuals involved in the conflict but also the integrity of the ecosystems which are its theaters.

It should be apparent to the reader by now that what we have called the 'environment and security nexus' is not a neat and clear set of issues. Rather, it is a collection of multiple forces that shape the relationship between human security and environmental integrity at any given place and at any given moment in time. Previous sections in this chapter have attempted to lay out the context of our study as well as to provide a snapshot of the various aspects of these forces, ranging from ecological endowments to institutional structures and developmental constraints. We will now seek to better understand the nature of the environment and security nexus in Sri Lanka by studying a set of actual cases that highlight how these forces have been implemented in the contexts described above. It should noted that, this being a new field of study, and in the absence of a significant literature on the subject in Sri Lanka, the ideas presented here are necessarily based on the author's own experience and interpretations of the following cases.

Throughout the process of developing these ideas, the key question was: "When does an environmental issue become an environmental security concern?" The answer, of course, depends on the security indicators that become apparent in a given environmental issue. At a minimum such indicators of security, viewed through the conventional security window, would include the presence or involvement of police and security personnel, either directly or indirectly. Thus the appearance of police at a protest march or campaign activity on the environment might indicate that the issue is tilting toward becoming one of environmental security. Similarly, if any other action requires reference to security clearances or needs security approval, one could infer that it has security relevance. At a different level, however, the very security of individuals may be threatened even if organized security forces are not involved. Environmental calamities, therefore, could be seen to jeopardize the human security of those involved. Without necessarily making a judgment on which particular indicators of security are the most valid, the following sections provide a sampling of case studies from Sri Lanka that highlight the various stresses on its environment and security nexus.

Case Study: The Accelerated Mahaweli Project

The Mahaweli Ganga is the longest river in Sri Lanka. Its source is in the western Sri Pada Hills, and it runs across the country, flowing east and then swinging north to empty into the sea at the Bay of Trincomalee on the northeastern coast. Several ruins of anicut-based (wier) constructions at Elahera, Pabbatanta, Angamedella, Minipe, Kalinga, and Nalanda Oya contain evidence that the river has been a major source of irrigation for many, many years. Indeed, some of these irrigation structures, such as the Minipe anicut, have been resurrected for present-day use. The idea that the river could be utilized for the purpose of irrigation is, therefore, neither new nor novel. Indeed, plans to build modern irrigation structures along the river have been in the works for half a century and a feasibility study was undertaken in 1958. It came up with a proposal to divert water from the river near Kandy to irrigate fields in the north-central province and also to generate electricity. This proposal, however, was heavily criticized and rejected by the government.

In 1963, the government commissioned an extensive study in collaboration with the United Nations Development Programme (UNDP) that came to be referred to as the Mahaweli Master Plan.

While this document has since undergone various transformations, it remained the basis of all subsequent discussions on the subject. The Mahaweli Program as outlined in the Master Plan was unprecedented in both its magnitude and its complexity. This program was followed by a set of phased implementation plans, the first of which included three projects. Implementation began in 1970; by 1976 two diversions had been completed at Polgolla and Bowatanne and about 500 families had settled in the north-central province.

The direction of implementation changed after the new government came into power in 1977. It decided to accelerate the process and telescope the projects for completion within six years. This new thrust in implementation became known as the Accelerated Mahaweli Development Programme (AMDP). For the first time in the history of a development project in Sri Lanka, an Environment Impact Assessment (EIA) was conducted. It highlighted the various environmental impacts of the project and proposed mitigation measures. These were instituted in an Environmental Action Plan (EAP) to be implemented alongside the AMDP. The recommendations of this plan included the establishment of an Environmental Action Committee (EAC) that was to meet regularly and monitor the progress of the project and of the mitigation measures. The author was a member of this action committee and all reflections are based on that experience.

Although they were not recognized as such – and certainly not categorized as such – many environmental security issues were raised during this implementation process. Decisions relating to the resettlement of affected persons and processes of providing them with assistance for resettlement were invariably accompanied by uncertainty and trauma. These decisions, of course, lie at the very center of what we have called the environment and security nexus. The uncertainty, fear and shock that accompanied having to move to sometimes harsh new habitats, which were completely different from their earlier abodes, were immense. Some, however, were able to overcome the inherent insecurity that came from being suddenly thrown into unfamiliar settings to become successful in their new habitats, while others continued to struggle against the odds.

Except in terms of hydroelectric power generation, the project has not been able to accomplish its economic objectives. While some have obviously benefited from it, the promised socio-economic benefits have not been delivered. Indeed, for some people, life has become harsher rather than better because of the project. The trumpeted benefits of modern irrigated agriculture proved to be exaggerated, especially since farmers were unable to cope with rising costs of chemical fertilizers, pesticides and mechanization necessary for large-scale agriculture. In 1993, for the first time in Sri Lanka, farmer suicides were recorded in the Mahaweli area.

Should this be considered an example of increased environmental insecurity? While there is no clear answer to this question, one can say that for a larger population than anticipated, and in ways that were totally unanticipated, the project caused increases in insecurity. On the other hand, however, one might also claim that the project had significant environmental benefits. In terms of energy, for example, it has allowed Sri Lanka to follow the path of cleaner hydroelectric power rather than that of more polluting thermal options. A conceptual lesson from this case is that the very same activity can make some people feel more environmentally secure while simultaneously making the lives of others less secure.

Case Study: *The Veddha Community Issue*

Related to the Mahaweli project, one environmental issue that remained contentious until very recently was the impact of the resettlement of the indigenous Veddha people. They are a small tribe whose livelihood was based on essentially hunter-gatherer methods. As the catchments for the reservoirs were designated, the area where the Veddha people lived was proclaimed a protected national park. Sri Lankan law does not allow any human settlements in national parks. This meant that the entire Veddha community had to be resettled.

Figure 5.1
Protected areas and Veddha areas

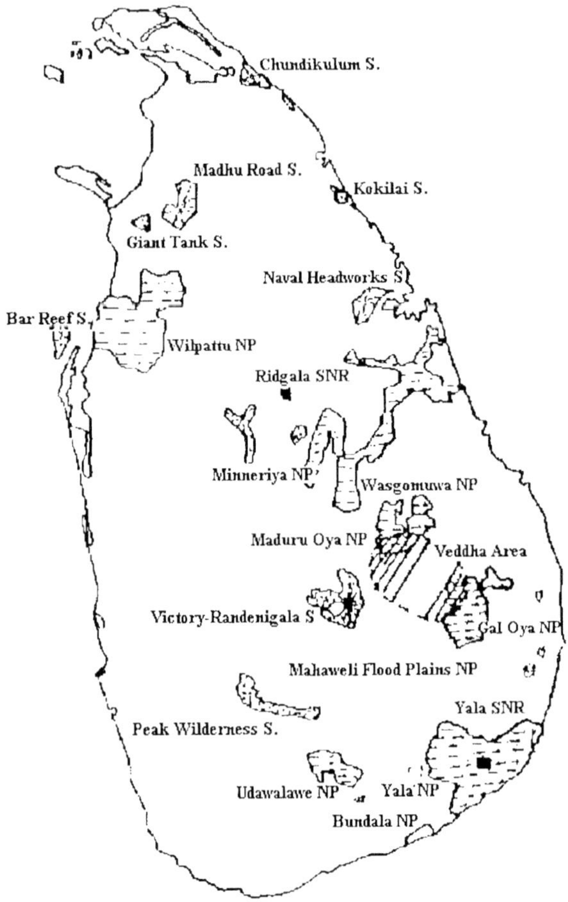

The development authorities took the view that these people should be brought into the mainstream; that they should be resettled in developed areas and given training and opportunities for commercial agriculture. The EIA team felt that the catchments declared as national parks could, in fact, incorporate the Veddha community since their lifestyles were already in harmony with their environment and the impacts that a mainstream settlement might have had in a protected area did not apply to this particular community. Their suggestions were overridden by the drive to bring the Veddhas into the mainstream of Sri Lankan culture. The development thrust prevailed and the Veddhas were resettled as one community in a new area, Henanegala. However, despite all the promises given by the authorities, a sector of the Veddha people, including their chief, resented and opposed the resettlement. This situation not only caused ill-will between the community and the authorities but created a serious rift within the community itself. At the end, all but seven families closely related to the chief left their land for the new agricultural settlement.

The lives of the newly-settled Veddhas were quite miserable during the initial stages. A special officer had been assigned to the community in order to help them in the transition from a hunting-gathering lifestyle to settled commercial agriculture. The social turmoil was considerable. The older generation, in particular, had difficulty adjusting to this new way of life. Since the settlement was established in 1984, there have been many instances of conflict with authorities; these lie very much at the nexus of environment and security concerns for the Veddhas. First they found themselves expelled from an environment that they fully understood and were comfortable with. Then their livelihood and security were imperiled as a whole set of skills at which they excelled became redundant and were replaced by a new set of skills with which they were unfamiliar. The ensuing conflicts often needed the intervention of a government official to resolve issues between the Veddhas and the authorities. Over time, the younger Veddhas have begun to adjust to their new surroundings at the cost of losing touch with their traditions; the older folk continue to resent the forced move.

The seven families that remained in their original homesteads were in clear and open violation of national park regulations and in full conflict with the authorities. With the help of outside sympathizers, they took their case to court. In 1988 the chief, with assistance from human rights lawyers, filed a criminal case against the then Director of Wildlife Conservation. Many incidents of direct conflict between the authorities and the group were recorded and often had to be settled by higher authorities. The court cases remained unresolved until 1990 when the author, as the new Director of Wildlife Conservation and a member of the original EIA team, decided to seek a different approach to settle the issue. With the full backing of the president, a program was proposed which would deregulate part of the national park for use by the Veddhas. This was initially accepted and, as part of the agreement, the court case was withdrawn. However, the relationship between the authorities and the Veddhas has remained generally antagonistic; this is a case where environmental and human security concerns seem to be at odds with each other in a complex and intertwined manner.

The legal rights of the Veddhas were finally decided in 1998; granted a formal recognition of their lifestyle and permission to carry on with their activities, provided they did not cause harm to the ecosystem. The situation has since become less hostile, although the wildlife authorities remain unhappy. The cost for the Veddhas has been high and one cannot be sure if there has been much gain for the environment either. The result has been a loss of environment-friendly indigenous traditions. Although the seven families, and particularly the chief, fought their way through using sheer dedication and grit, their ultimate victory is itself tenuous – indeed, it is not that they have gained anything from the episode, it is only that they have not lost as much as they might have. More importantly, the episode highlighted a false conflict between the environment and human security and in pitting the two against one another left both much bruised.

Case Study: The Human-Elephant Conflict

Another more persistent issue in Mahaweli area is the so-called 'human-elephant' conflict. Elephants that once roamed wild in the forests of the Mahaweli area have been forced by humans into restricted protected areas and diffused scrub patches. The need to conserve an area for environmental amelioration compared to the need for vast open tracts of land to serve as refuges for the displaced animals was one of the concerns of the Mahaweli EIA report. A fairly large area, about 46 percent of the entire tract, was eventually earmarked for the elephants. But this action of conservation has had its costs as well.

Figure 5.2
Human-elephant conflict areas and location of electric fences

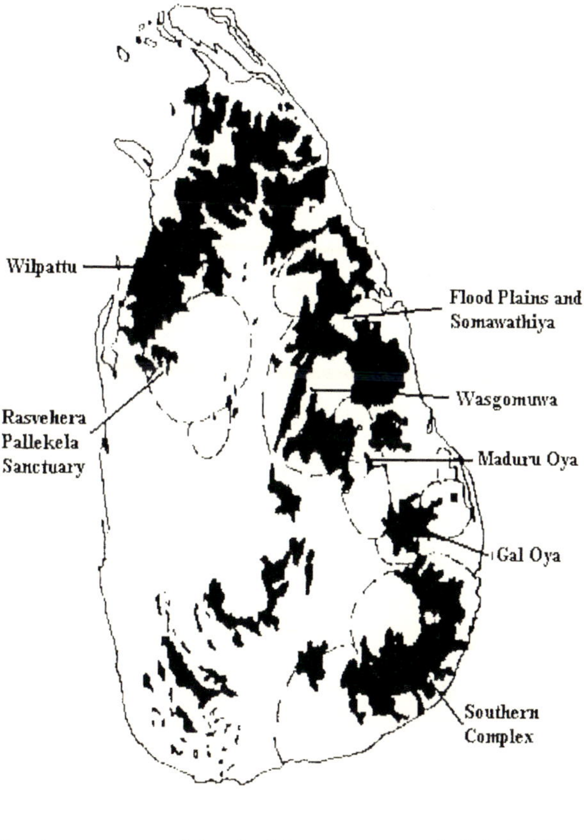

The elephants, unaware of the decisions made on their behalf, tend to wander beyond the designated area, causing considerable damage to crops and occasionally to property and life. Indeed, a new conflict has been created between the continued security of the elephants and that of their human neighbors. This has taken on fairly serious proportions. In some situations the elephants' wanderings have resulted in the complete cessation of movement of people. In one area, the lighting of lamps in houses became an invitation for elephant attacks. In retaliation, some communities have become openly hostile towards the elephants, leading to a breach in law and order. This is a clear case of imbalance between environmental and human security.

This issue, however, is not confined to the Mahaweli area but has spread over almost all of the dry zone. The extent of the problem is reflected by the statistics of elephant and human deaths. In 2000 it was officially recorded that 148 elephants died due to anthropomorphic reasons, as opposed to only about 27 natural deaths. The number of unrecorded deaths will most likely make the actual death toll higher. The human cost has been estimated to be at least 35 deaths, primarily farmers in lands adjoining the elephant habitats. The human security implications of this problem, although often local, are serious. Indeed, the human inhabitants of these regions now view their elephant neighbors as a security threat to their own lives and livelihoods and therefore, undertake actions that threaten the security of the elephants.

Attempts at resolving the conflict have included relocating animals identified as crop-raiders, driving elephant herds deeper into the forests, and building electric fences. Communities have created participation programs to maintain these fences. However, the cost for electric fencing and constant relocation has become a heavy economic burden, and the problem remains persistent and acute. (See Figure 5.2).

Case Study: The Kandalama Hotel Issue

In 1990, a leading Sri Lankan tourist organization decided to build a new mega-hotel. The location selected for this venture, Dambulla, was in the heart of Sri Lanka's cultural triangle. While the purpose and need for a new hotel was quite within the established development needs of the country, the proposed site became a hotly debated issue that pitted development needs against environmental constraints.

Environmentalists raised a host of concerns about the selected site, including issues about the social and cultural implications of such a development. The proposed site of the Kandalama Hotel was in the immediate catchment of a major reservoir and in an area that boasted historical significance dating back 2,500 years. In the midst of strong protests, an EIA was conducted, the results of which were highly criticized because it had failed to incorporate any analysis of the area's cultural dimensions. Charges of political misconduct were also levied because the project violated existing irrigation laws. Against this backdrop there began a massive protest organized by a coalition of environmentalists, religious organizations and concerned citizens.

State security forces confronted the protestors, an action which was tainted by political overtones, as the government was visibly backing the construction project. Feelings of hostility ran high and at times sparked significant confrontations among various groups. Ultimately, the construction site had to be totally closed to the public during the project development phase. In the middle of this uneasy situation, the hotel was eventually completed.

In this case, the environment and security nexus seems to have entangled itself in both violence and animosity. A monstrous concrete hotel stands at the disputed site as a monument 'eco-resort' and an environment-friendly facility. While it may eventually become so, its history is forever soiled by its past. The environmental protests have since subsided, but memories of the conflict linger.

Case Study: The Coal Power Plant Program

Our next case study relates to the energy sector. Since the 1980s, Sri Lanka has unsuccessfully tried to promote the construction of a coal-powered electricity generation plant. The first attempt at Trincomale in 1985 failed to meet the technical criteria and was rejected for want of a favorable EIA. A second attempt in 1989 to construct a similar facility in the south was unsuccessful for political reasons. The inhabitants of that region very clearly denounced the erection of a coal-driven power plant, sending a clear message of antagonism to the ruling political party, which in turn defused the situation by essentially shelving the project.

A third attempt was made in 1998 by planning construction along the west coast at a place called Norchcholai, close to Kalpitiya. In this case, the work actually began, but not before a protestor lost his life at the hands of the police. Although an EIA has been completed for the power plant, protestors remain opposed to its construction for a variety of environmental and social reasons. The result has been the suspension of project implementation. Indeed, the religious leadership which makes up part of the protest coalition has convinced the president of Sri Lanka to promise that the project will not proceed if the people disapprove of it. However, the Minister for Power insists he will go ahead with the project in spite of this fact. The situation has become quite volatile. The level of antagonism is so high that even though the project is virtually suspended, state officials from certain agencies are not welcome in the Norchcholai area.

Multiple securities seem to be at risk in this case: the nation's energy security, the environmental security of the location, the livelihood security of those who are economically affected, and the very personal security of state officials assigned to work in the area. This case highlights the complex nature of the environment and security nexus.

Case Study: The Eppawala Phosphate Mining Issue

Our final case study relates to a mining issue. The Eppawala mineral deposit is a rock phosphate deposit of very high quality, located in the north central province at Eppawala. While it is believed that the deposit also contains rare earth's, their precise extent and quantity is yet to be properly ascertained. According to crude estimates, the deposit is thought to cover an area of about 55 square kilometers. This area is however home to numerous traditional villages and archeological sites.

The main opposition to the project is concerned with the loss of these traditional homelands and the livelihoods of their inhabitants, as well as residual environmental impacts of strip mining. Finally, the involvement of a foreign private company has raised concerns about the loss of national sovereignty.

Protesters claim that the mining project will have devastating environmental impacts and that the economics of it are dubious and not necessarily in Sri Lanka's best interests. Agitation around the project has already resulted in large social polarizations leading to threats of bloodshed. Indeed, some residents of the area have threatened a 'sit-in-to-death'. The immediate security of the location and its inhabitants has been clearly and severely threatened. In a recent and unprecedented judgment, the Sri Lankan Supreme Court decided in favor of the protestors and has effectively stalled the project for the time being. However, given the potential of the site, the issue may well raise its head again. The point of contention was not merely the interests of a certain community versus those of corporations, but the larger issue of balancing the need for development with the necessity to maintain environmental quality. Once again, it is the quest for this balance that lies at the heart of the environment and security nexus.

Final Thoughts

The analysis of the various case studies described above points toward some interesting conclusions. Most importantly, all conclusions highlight the evolving nature of the environment-security nexus in Sri Lanka. Based on the author's own experience of over thirty years in the Sri Lankan environmental movement, it is obvious that individual responses to environmental issues have changed significantly. In essence, environmental concerns have not only become more prevalent and more mainstream, but they have also become more security-oriented. Environmental activism in the past was largely tied to more technical matters and operated using the conventional means of political appeals and protests. Environmental activism today is often linked to larger processes of social mobilization and can take on the form of mass movements. It is usually intertwined with other concerns of human security, taking part in a larger agenda of social change of which the environment is only a portion. Unfortunately, these protests often lead to violence or threats of violence. The environment is no longer a concern that can be understood on its own terms alone. It is best contextualized and properly understood within the larger dynamics of human security; as one element in the melee of forces that make the lives of citizens in a polity more or less secure.

As a final thought, it should be noted that in addition to the various physical aspects of the environmental concerns discussed above, the impacts of changing lifestyles, development processes and cultural priorities must also be taken into account. On the one hand these changes directly and economically impact human well-being. On the other hand, cultural traditions themselves are in flux and there exists a growing concern that Sri Lankan society has become more violent, intolerant and abusive than ever before. Whether or not this is a direct result of modernity, economic development and globalization, the fact remains that the face of society has changed in the last half century, and will continue to change faster than ever before. The environment and security nexus is centered, therefore, not only on the conflict between human economic security and ecological integrity, but is also inexorably tied to the cultural robustness of Sri Lankan society and its institutions.

Chapter Six

A Cultural Theory Perspective on Environment & Security in Nepal

AJAYA DIXIT
DIPAK GYAWALI

ON 9 MAY 2000, Gajendra Sahi, the Nepali Congress Chairman of the Dandeldhura District Development Committee, led a delegation from his district to meet Prime Minister Girija Prasad Koirala. The committee chairman handed over a protest note opposing the implementation of the Poornagiri Dam, whose construction downstream of the Pancheswar Dam was proposed under the Integrated Treaty on the Mahakali River. Earlier, the District Committee of the Nepal Communist Party United Marxist-Leninists (UML) had passed a similar resolution opposing the Poornagiri Dam proposal, because the dam and the reservoir would submerge the highly productive valley there. The Poornagiri is a re-regulating dam that would be built downstream from the 315-meter high Pancheswar Multi-Purpose Dam. The Pancheswar Dam would also submerge a large land area of the Dandeldhura and Baitadi districts. According to preliminary estimates, about sixty thousand people would be displaced. For its part, the proposed Poornagiri Dam would submerge a substantial amount of cultivated land in the Dandeldhura District.

The story of water resources, as exemplified by these dams and the struggles associated with them, is a story of how conflicts over environmental issues become intrinsically intertwined with the battle for human security, particularly in developing countries such as Nepal.

The Mahakali Treaty: Unmet Expectations

The Mahakali Treaty between India and Nepal was ratified by the joint session of the Nepali Parliament on September 20, 1996 with a two-thirds majority that included the then-ruling Nepali Congress and the opposing UML. This Treaty aimed to set the stage for the development of the Pancheswar Multi-Purpose Project. The Mahakali Treaty, which overrode the earlier Memorandum of Understanding on the Tanakpur barrage, stipulated specific provisions regarding the ways in which India and Nepal would share the water of the Mahakali River and the energy that the high dam would generate. In order to derive maximum benefit from the project, the Treaty laid out the following provisions:

- The Detailed Project Report (DPR) to be finalized by both countries within six months of the Treaty's coming into force. (Ratification by Nepali parliament; India's parliament does not have to ratify decisions made by the government).
- An agreement for the financing and implementation of the project, including the proposed establishment of the Pancheswar

Development Authority, to be negotiated and finalized by both countries within one year of finalization of the DPR.

- The project to be completed within eight years of the date of the agreement, subject to the provision of the DPR.

Today, five years after its ratification amidst much hype and hoopla, the Treaty's basic provisions remain unimplemented. Even the DPR remains in limbo. Moreover, the likelihood of its ever being implemented in its original spirit is now in question.

What went wrong?

While ratifying the Treaty, its proponents had argued that it would make Nepal and its people more secure by generating revenue and other benefits from storing large volumes of water: irrigation, power, recreation, navigation fishery development, etc. However, on the insistence of the project's critics, four stricture clauses redefining the provisions of the Treaty were also passed prior to ratification. These clauses related to water rights, the price of electricity, the status of the river and disagreements about the origin of the river at Kalapani. The disputes over these issues still remain unresolved. Because of these lingering differences and the problems of arriving at a mutually acceptable DPR, the opponents of the enterprise now consider it to be an illegitimate venture despite the two-thirds majority which originally ratified the Treaty (Gyawali and Dixit, 2000).

From the side of the Nepali Congress, the former Minister of State for Water Resources and the current coordinator of the Nepali Congress' water resources policy has said that until the issue of Kalapani and the source of Mahakali is settled, it is impossible to prepare the DPR (detailed project report) of Pancheswar. He goes on to point out that if India agrees to Nepal's position on Kalapani, the implementation of the Mahakali Treaty may move forward, but even then it was doubtful if the DPR could solve the issue of water sharing in an agreeable manner. Indeed, this leaves the Treaty frozen in a deadlock (*Samanata*, 2000; Ghimire, 2000).

The Mahakali Treaty is a classic example of a bureaucratic response that has filtered out the social and political questions from a resource pool dominated by the engineering mindset. The hierarchic solidarity in both Nepal and India concurs with this mindset and the voices that have sought to highlight alternative points of view have been both marginalized and ignored. Criticism came from social activists who highlighted the social, economic and political risks of the chosen path. Many of the problems, they insist, need political rather than engineering responses because they relate to the rights, obligations and issues of territory that an engineering document cannot resolve.

Even if the differences are resolved, the implementation of a regional project of this scale will be rife with uncertainties and disputes. The example of the Tehri Dam in Garhwal (Uttaranchal, India), located 150 kilometers due west of Pancheswar, is a case in point (Aryal, 1995). The Tehri project on the Bhagirathi, which was sanctioned in the late 1960s, still awaits completion. The three-decades-long debate on that project highlights the enormity of the sense of insecurity that can arise between parties by the loss of sustenance at the hands of large bureaucracies running mega-projects from afar.

Not only have the existing arguments over the compensation and rehabilitation of displaced families not been resolved but the possibility – or rather probability – of future disputes becomes ever stronger. As they looked at the case of neighboring Tehri, the reality of having to leave one's own home began to dawn on those Nepalese that could be affected by the construction of this dam. Their collective stress expressed itself in the form of a memorandum to the government from the district committees of the two main political parties of Dandeldhura District. What is ironic in this protest memorandum is that the committees represent at the grassroots level the parties whose central leadership overwhelmingly ratified the Mahakali Treaty at its inception. It shows that the perception of security can differ dramatically as the scale changes, even within a political party.

Security & Risk

The point underscored by these disputes over large projects is that security is not simply the desire of the government to provide for its citizens (Thompson, 1997). A proposal made by the state with the aim of providing energy or water security to some people is seen by others as a threat to their own livelihood security. People living in the paths of proposed reservoirs see an immediate security threat in the possibility of having their land drowned. Such threats, however, are often ignored in the hierarchic rationality of the engineering mindset. When those parties who feel threatened join forces to reject the implied rationality, the result is often an impasse.

The case of the Sardar Sarovar Project in Narmada (India) typifies the nature of opposing rationalities on a confrontational course. Supporters of the dam outline the benefits of the project while acknowledging the cost, which they argue is resolvable by procedures. They point out that the needs of the 40 million people who are expected to benefit from the project (those whose security will be enhanced because of the dam) dwarfs those of the 240,000 people who will be harmed by the project (Postel, 1999). The Narmada Bachao Andolan (NBA), the major civil society coalition of project stakeholders, rejects this notion of security and instead argues that the construction of the dam must be halted in order to avoid the insecurity that the displaced people would face. These examples of mega-projects highlight the dynamic notion of 'securities in conflict', thereby providing a different lens through which to view the issues at stake.

The notion of 'security' is intricately related to the issues of survival and livelihood. Therefore, access to water and energy become central in how these issues are articulated in the lives of the Nepali people. Ways in which they cope with insecurity is a function of how they perceive risks. It is useful, therefore, to understand how people perceive and deal with issues related to risks. An individual makes choices, forms alliances, and takes part in a community where his/her relationships are defined. Social scientists who study risk allow that individuals are active organizers of their own perceptions and how much they choose to fear, preconditioned by their need to support their own way of life. These ways of life have four basic groupings, or 'cultural biases', in institutions that can be characterized as hierarchical, egalitarian, entrepreneurial, or fatalist. Each has a different institutional filter, through which risks are perceived and assessed. Risks are akin to a lens for sharpening the focus on social organization itself (Douglas, 1985).

Hierarchies (or governmental departments) tend to risk minimization: they consistently make it clear to their members that they can best secure their interests through conformity. Those who see rewards in taking risks espouse an entrepreneurial (or individualistic) outlook. They gravitate toward an adversarial, or heroic stance that requires challenging the real distribution of power, whether over scientific data and interpretation (as in the case of independent academics) or dissemination of information (in the case of investigative journalists).

The entrepreneurial urge could manifest itself in challenging the official line as well as in manipulating the official machinery into a position from which it could benefit (in the case of consultants and CEOs). Much of the entrepreneurial urge is, however, expressed in trying to devise clever solutions and profit from them. It is this inherent suspicion of conformity and individual profiteering that galvanizes volunteers with an egalitarian and high risk-averse outlook, "an association whose members are able to boast more of having kept together than of having achieved anything in particular" (Douglas, 1985) Fatalists, unlike the others, are passive about risk. They are so attached to the external order that the best they can do is to cope with, and, at worst, react in anger against a situation in which they find themselves. These are the street people whose status is hard to change and who lack a clear sense of group cohesion. They fear uncertainty and look toward authority to provide predictability in their lives.

Water-Related Livelihoods & Security

Using the conceptual constructs about the nature of security and risk introduced in the previous section, let us now apply these insights to the problematique of livelihoods and security in the water-related sectors. Human security is closely related to issues of livelihood, which in turn are dependent on specific activities like irrigation, food and drinking water supply, pollution, flood mitigation and energy. We will now turn our focus to these activities.

Irrigation

The modern irrigation system led to a political bureaucracy centered on the construction of new institutions to increase food production. This approach not only discounted areas irrigated by systems built and managed by farmers, but also assumed that if water were delivered to the canal, productivity would automatically be increased. However, the question of irrigation lies on the line between the private and the collective; its impacts are neither linear nor deterministic. Agricultural land is privately owned, and its management is based on the judgment of individual farmers and by their interaction with one another in the realm beyond their households and villages. Irrigation introduces collective values into individualistic patterns of organization because it contains a common resource pool -- flowing water. Farmer-managed irrigation systems have evolved over time in order to best share the resource. Their systems include sanctions like fines and decrees that stipulate when groups must unite in order to clear the canal or rebuild the *syauli* (brushwood) dam after a monsoon flood washes it away. They have acquired a certain irrigation water to entire communities.

The situation, however, is not as encouraging in government-built systems where reliability of and access to water supplies are often unsatisfactory. Because these state-run systems are often unable to provide easy access to water, they are being handed over to groups of farmers for management. This approach uses lessons learned from farmer-managed irrigation systems while also taking advantage of collective management. The farmers are allowed limited access to the canal network while the governing agency maintains control of it. The imbalances in and poor reliability of water levels, coupled with the invention of a pump whose use is backed at the highest level by an incentive regime, has resulted in large-scale disealisation of irrigated agriculture. A pump is flexible and reliable, and it makes a farmer less dependent on other sources for irrigation. While the use of groundwater irrigation has contributed to an increase in food production, its negative impacts include groundwater depletion and equity problems faced by latecomers to the system. This practice has caused increased pollution in many cases by allowing saline water to intrude into fresh water aquifers in both coastal and arid zones.

Degradation and Pollution

Rising population and urbanization are leading to increasing stress between the traditional and statutory water management organizations. Even though this shift in population would require some reallocation of water from the rural agricultural sector to urban areas, this is a trend that is likely to continue because of economic liberalization policies. Industrialization has pollution as its by-product; it also affects the quantitative allocation and sharing of water in the dry season.

In both types of systems, urbanization, pollution and the common preference to opt out of a collectivist system often cause problems. While the farmer-based system may have some social legitimacy, it lacks the ability to react to new kinds of stress. Pollution upstream would render the available water unusable and externalize the cost of treatment, which can also create disputes between the upstream and downstream users who depend on the same water source for their very existence.

Floods

Flooding often occurs in the region. Unfortunately, the governmental response to disasters focuses principally on the natural or biological events which trigger them (Blaikie et al., 1994). Indeed, the phrases 'natural hazard' and 'natural disaster' are often used interchangeably. Policymakers, donors, and relief and development agencies treat disasters as isolated events that break the continuity of the 'normal' way of life. This approach relies upon the assumption that these events are departures from 'normal' social functioning and that recovery from disaster means a return to normal conditions. As a result, interventions to mitigate disasters are *post hoc* responses made under the assumption that an emergency support in the form of relief will change the situation of hardship; they do not solve the underlying problems.

Recent experience shows that disasters, and the environmental stress and human insecurity that accompany them, are the outcome not only of natural hazards, but also of socio-economic structures and political processes which actually make individuals and families more vulnerable to disasters. In other words, various sub-systems are at work where nature manifests herself very differently. This perspective focuses on the ways in which these social systems conduct themselves. It begins with the belief that natural hazards are significant and normal characteristics of the regions where they occur most often, and goes on to recognize that vulnerability is equally important. Vulnerability is a characteristic of a person or a group which measures the capacity to anticipate, cope with, resist, and recover from the impact of a natural hazard (Blaikie et al., 1994). It can also be defined as social resilience in facing a hazard.

In Nepal the 'normal' daily life of most of the population is vulnerable. Individuals and families in many communities find it hardest to reconstruct their livelihoods after a disaster strikes. These people do not live under vulnerable conditions out of ignorance. Instead, most have little freedom to choose where or how they live. Low-income families, for example, often have no other housing options except those in vulnerable locations, such as flood plains. They are forced to live there not because land-use planning is poor, but because the prevailing agrarian relations and the practice of social and economic exclusion in their society make them unable to own property in safer areas.

The focus on a broader political and economic outlook also needs to recognize the specific details of each disaster. Only a combined analysis of natural events, technology and societal context can provide a realistic understanding of floods and their impact. Extreme natural events, like cloudbursts that can bring more than 500 millimeters of rain in 24 hours, are common here. Every perspective on natural events, technology and society is necessary, but none by itself is sufficient; we must separate them in analysis but integrate them in policy formulation. Through such distinctions, we begin to analyze natural events as well as processes that make individuals and families vulnerable.

We capture the interactions among various sub-systems and locate the constraints among them in order to be able to design responses that are specifically tailored to a specific context (Moench, et al., 1999). Instead of focusing on technical solutions such as embankments or dams, the insurance schemes, primary health services, and education should be prepared to face the impact of floods at the individual, family and community levels. Such 'risk-resilience' approaches that are better suited to respond to the human security concerns posed by floods.

Energy

Among the most contested issues in Nepal is that of hydropower development, which had until recently operated under an unchallenged monopoly run by the government. This, of course, has left direct and profound impressions on livelihoods and human security. Hydropower came to Nepal in 1911, about two decades after Nikola Tesla patented the commercial production of the current alternating motor. Ninety years later, only 15 percent of the population has access to the electricity supply from the national grid and hydropower accounts for a scant one percent of the nation's energy use.

Moreover, Nepali consumers pay perhaps the highest price per unit of electricity in the world, a fact which was acknowledged by Deputy Prime Minister Poudel at the beginning of the general meeting of the Langjung Electricity Development Company (LEDCO) (Kantipur, 2000). The result is a stunted possibility of developing electricity as a means of production, thereby facilitating a transformation of the nation's agrarian economy into one based on industrial production. At the same time, due to the fact that local capacities to plan and implement projects are still in their nascent stages, the backward linkage of the investment made in developing hydropower projects remains weak. A preliminary analysis of the investment that has gone into the development of hydropower projects after 1995 indicates that less than five percent returns to the local economy or results in the creation of jobs (Panday, 1996; Dixit, 2000). The rest is used to purchase goods and services from outside the country.

When a new government took charge in 1991, one of the main issues it faced was the Arun III hydropower project. It was supposed to be at the end of a road 120 kilometers long, which had yet to be built. The size of the project was eventually downscaled, but the budget was estimated at more than one billion dollars. Questions were raised about reductions its detrimental effects on local capabilities as well as the unbearable risks involved in the project. Not only did egalitarians critique the project, but in this case they also provided alternatives that were cheaper and would produce more power quicker.

In 1995 the World Bank, the main investor in the project, withdrew its support. In an overt sense, the campaign that attacked Arun III in Nepal would be the first example of this in post-1990 Nepal, when social auditors were in action. The campaign against the Arun III brought academics and activists together to challenge a project about four times more expensive to build than what the private sector was able to do, and to also present a set of alternatives, which would be cheaper, more effective and more efficient than the proposed project. After the cancellation of the Arun III, six projects quickly began construction. These are expected to produce a third more electricity than the Arun III would have, at approximately half the cost as well as in half the time (Panday, 1996).

After the Arun III fiasco, two distinct paths have formed for hydropower development in Nepal. On the one hand, government-led initiatives continue. On the other hand, there is a spurt of initiatives wherein individuals and district representatives have formed cooperatives to develop hydropower projects using local resources, investments and skills. Notable among these are the Lamjung initiative and the Piluwa Khjola Hydropower Company, which have mobilized local investments and explored ways to minimize costs. This approach is also guided by concerns about extending services to areas that do not have hydropower available to them. Not only do these initiatives champion the self-help hydropower development mode, but they also question the large-scale projects proposed to export power to India. These two approaches represent dramatically different biases. They also point toward a different approach to resolving the 'securities in conflict' – one that involves different actors and arrives at different answers than the traditional hierarchical approach does.

Security & Solidarity

In Nepal, natural disasters interact with societal vulnerability, disruptive development interventions, and perverse market exchanges to create an atmosphere of insecurity. Water pollution, sedimentation in and shifting of river channels, land rights, tenancy conflicts, a lack of local maintenance capacity, high electricity costs, dependence on external sources for developmental needs and many other issues add to this harsh reality and put further stress on the system.

What does the future hold for human security, when it is related to water access for irrigation, drinking, and respite from flooding and pollution including energy needs? This is an important question in a milieu which is marked by a complex interaction of physical, social, cultural and institutional factors. As in other nations, individuals, families, and communities in Nepal perceive security differently and respond differently to challenges. The idea of a single 'Nepali perspective' on security would therefore be a misnomer. Instead, this paper suggests three perspectives that result from the conceptual discussion of security and risk presented earlier:

- Hierarchies would seek additional security through procedural rationality; their technological and developmental choices are biased toward the large scale and therefore require additional expertise and control.
- Those in the individualistic mode would seek security through innovation and networking; their choices veer toward the path that provides the most substantive benefits.
- Egalitarians will seek social justice as the foundation of human security; they are suspicious of choices not easily controlled by the community.

Security enhancement therefore must occur within the constraints imposed by the region's climate, hydro-geology, and geo-tectonics, including financial macroeconomics. It must also match the diversity of its social base. Hydro-geology and geo-tectonics are the contexts within which possible solutions must be negotiated. These constraints are real and cannot be overcome. Because variability among natural, social and institutional factors is high, the response must also be varied – several problems require several solutions.

The most common practice in all water-related areas is to define the problem in a unitary mode and generally seek one solution within a hierarchic framework: a large dam, for example. Historically, however, security issues have been viewed through state-centric lenses without taking a single cue from the grassroots. With the emergence of new stress points within and between systems, the unitary approach has become a recipe for depletion of social resilience and, thus, an increase in human insecurity.

Institutional pluralism doesn't just mean organizational multiplicity, with ministries, departments and parastatal organizations multiplying and cloning themselves at various levels. Security enhancement via access to water and energy supplies has to follow the mutually reinforcing homeostasis of state, market and civil society. The market, by its very nature, would affect change, and the state would provide a regulatory framework to ensure a fair playing field. A cautionary civil society such as that of the social auditors, much like the third leg of a stool, would provide a countervailing check on the rapacity of the market and the powers of government. So far, there has existed an uncontested terrain (Thompson, 1997) where the state has been playing the role of both the market and civil society, essentially stunting the evolution of either a competitive market or a vibrant civil society. The result is a weak market and an uncommitted civil society juxtaposed with a rent-seeking state.

The status quo must be challenged, even though to do so is an action uncomfortable to those accustomed to having free rein of an uncontested terrain. Contestation and cautionary voices can sometimes lead to benefits, as was made obvious by the campaign against the Arun III. The cancellation of its funding by the World Bank led to the initiation of a more pluralistic path to hydropower development, though at this point in time less coordinated questions are being asked about the high cost of electricity and its links with the local economy. The same is also true of projects like the Pancheswar High Dam.

Compromises made in achieving the Mahakali Treaty, which was ratified without either the necessary homework and by filtering out questions about cost and risks, have had a severe backlash in Nepal. Political parties on both the left and the right have split, while the benefits promised appear to be further in the future than ever before; this has inflicted a deep sense of betrayal upon the population. Major political parties and the state structure in Nepal continue to get caught up in the ideology of unitary solutions, particularly in the form of building gigantic hydropower projects, as a salvation for the country's social and economic woes. Yet it is not at all clear whether, when and what these projects do for the human and environmental security of the Nepali people.

This mindset is reflected in a policy speech by Nepal's late King Birendra, given to the Nepali Parliament on 19 May 2000. In reaffirming the need to mobilize private and foreign investment in the development of water resources he said, "While proceeding with the Pancheswar Multi-Purpose project, the works relating to the preparation of the Detailed Project Report of projects such as the Sapta Koshi and Sun Koshi-Kamala Diversion shall be taken up." This fascination with the pursuit of technologically and organizationally inflexible options only serves to limit the scope for performing other necessary activities; in the process, the state ceases to become an institutional resource that helps its people enhance their security.

The key to meeting the future challenges of human security and their links to the environment lies in transcending the conventional way of looking at security itself. These securities can only be enhanced by resorting to institutional pluralism with an emphasis on the harmony between society and the environment and providing freer access to resources such as water and energy. The starting point for security enhancement is to build on the age-old resilience provided by families, neighborhoods, hamlets and communities as agents of security. Under such an approach the styles of management may vary at each level, and they are linked to practices of good governance at all levels, from the local to the national. This process consists of allowing space at all levels (from the village to the state and regional levels) for a genuine market that is open and competitive, egalitarian social auditors with transparent voluntarism, and honest regulation that is not driven by hubris or rent-seeking propensities on the part of the state.

In this process, the societal balance of power is key. As per the principle of balance of power, all three institutional sectors should be able to stand together in 'social solidarity'. In order for the sense of security to be enhanced, state, market and social auditors must exist in the system as a whole and must be provided with ample space for creative experimentation in the management of water, energy and risks. If the state-market-social auditor combination were actively engaged, a clumsy compromise could be found; it would be less than what each wanted but much more than what everyone would have if no compromise were reached. Constructive engagements between solidarities would ensure circles of improvement in technological choices and developmental security. A destructive engagement would result in impasse and stagnation if one solidarity (mainly hierarchic) made all the choices without consulting other voices. Security would also depend on ways in which far-sighted statesmanship could change the role of the state from a rent-seeker to a sagacious adjudicator. Such a 'three-legged stool' would bring greater security, not just to Nepal, but possibly to all societies at every level (Gyawali, 2001).

Chapter Seven

Energy & Security in South Asia: A Multiple Frameworks Analysis

KUMUDU GUNASEKERA
ADIL NAJAM

In the late 1940s, when much of the South Asian sub-continent achieved independence from British colonialism, conditions throughout the region — especially in political, economic, social and environmental settings — stood at relatively similar levels. While these countries still exhibit many important similarities, significant differences have emerged on particular development-related issues. One important area where South Asian countries have followed different paths, to different results, is that of energy.

This chapter seeks to understand how the 'energy paths' chosen by individual South Asian countries have differed; what the impacts are, if any, of these differences; what these countries might learn from each other; and what implications of the above impact on the energy security profile of the entire region.

The varying energy profiles in the SAARC (South Asian Association for Regional Cooperation) region are particularly intriguing, considering the similar historical and geographical backgrounds shared by these countries. Notwithstanding the differences in natural energy endowments of different nations, the institutional and historical similarities suggest that there is much room for shared learning among and between these countries.

This chapter follows the lead of the recent emphasis on broader definitions of 'security' — definitions which account not only for military and non-military threats to state security but also for internal and external threats to human security. We include, therefore, discussions of 'energy efficiency pathways' and 'energy impact pathways', which are internally-focused dimensions of energy security, as well as 'energy dependence pathways', which are focused externally. This allows us to adopt a maximalist understanding of 'energy security'; one that includes energy dependence (the traditional measure of energy security), energy efficiency (in economic terms) as well as energy impact (in environmental terms). In doing so, we place the concept of energy security in the broader context of sustainable development by imbedding it in still-larger developmental and environmental parameters in addition to traditional energy parameters. The use of multiple frameworks allows us to analyze energy security through three distinct, but relevant, lenses: those of economic efficiency, environmental impacts, and external dependence, respectively. Each framework focuses on one of these indicators to plot changes in the 'energy security' of individual countries. Inter- and intra-framework analysis enables us to compare and contrast these paths over time.

The chapter finds that in economic terms, no country has managed to develop much beyond energy subsistence, and barely that in most cases. Economic growth is nearly always accompanied by a corresponding increase in demand for energy. Since South Asia is still in the early stages of economic development and therefore has high growth rates, there is a much closer link between economic growth and energy demand. From a traditional energy security perspective, increasing populations, high economic growth rates, and transformations in the energy mix (from biomass to commercial energy) are likely to increase South Asian dependence on imported energy. This will render the region more vulnerable to fluctuations in global energy markets, which can weaken several forms of national security, including economic, political, social, and military security. From an ecological perspective, increases in energy consumption lead to higher emissions (sulfur, carbon, etc.), which consequently increase the potential for regional and global environmental problems. Interestingly, but not surprisingly, the preliminary evidence suggests that nations with stable (less volatile) political systems, liberalized economies, and insightful economic and energy planning have achieved the most progressive levels of 'energy security' in the region.

Energy Efficiency Paths (EEPs)

The first framework we use to chart the energy security profile of South Asia analyzes energy efficiency paths by plotting each country's energy intensity levels as a function of per capita income over time. Countries with higher per capita income levels and lower energy intensity levels are considered to have progressive energy efficiency levels.

The Energy-Economy-Development Link

Economic growth is a measure of changing income or wealth at both the micro and macro levels (e.g. GDP). Development, on the other hand, encompasses structural changes within a system (changes in infrastructure, resilience, empowerment etc.). Thus, economic growth correlates to development when it induces a structural change within a system. This implies that neither the numeric value assigned to growth nor its direction need necessarily correlate with the magnitude or direction of development. Indeed, economic growth is not the only component of development, although it is one that is widely pursued (Sen, 1984; Black ,1992; Banuri, et al., 1994).

On the macro level, energy services are essential for economic growth. No country has managed to develop much beyond a subsistence economy without ensuring at least minimum access to energy services for the majority of the population. On the micro level, energy increases the standard of living. It improves productivity, frees women and children from the daily grind of water and firewood collection, extends the workday, and helps communities interact with the rapidly globalizing world. Since economic growth is often evaluated in terms of welfare (utility) – measured as the willingness to pay for goods and services in relation to the ideal of Pareto optimality – it is appropriate to assume that energy demand would increase with rising incomes (Nakicenovic, et al., 1998). Thus, economic growth is nearly always accompanied by a corresponding increase in the demand for energy and visa versa. Furthermore, this link is stronger in developing countries than in developed countries (Munasinghe, 1995a).

Before the energy crisis of the 1970s, economic growth was nearly always accompanied by a corresponding increase in energy demand. However, as a direct result of the energy crisis (increased oil prices), the advanced industrialized economies managed to de-link economic growth from energy demand growth, primarily through increases in efficiency. They were able to reduce energy wasting and achieve better energy management techniques by restructuring and developing more energy-efficient technologies. Since developing countries are still in the early stages of economic development, have relatively high growth rates (for example, SAARC's growth rate in 1997 was 6.6 percent), and low energy efficiency levels, they have a much closer link between economic growth and energy demand (Munasinghe, 1995b).

Given this strong link between energy and the economy in South Asia, it is not surprising to find that the 1.4 billion people who live in the region attach a high priority to energy services. On average South Asians spend 12 percent of their disposable annual income on energy, more than five times the average for people living in industrialized countries (Munasinghe, 1995a; 1999a). It is apparent that energy services are high on the agenda; in economic terms, energy is a 'revealed preference' for people in South Asia. Thus, the energy sector in general and, more specifically, energy efficiency, are an appropriate surrogate measure for economic development and energy security.

The EEP Framework

The framework used in this section utilizes a widely used scatter plot time series model; it plots energy intensities as function of per capita incomes (Nakicenovic, et al., 1998; Goldberg and Reid, 1999). The energy intensity of a country is defined as the ratio of total primary energy supply (both commercial and non-commercial) (E) in tons, to gross domestic product (GDP) in US dollars, or E/GDP (in tons/ US dollars). This is plotted on the Y-axis. Energy intensities of a country change due to structural changes in the economy and energy, as well as changes in energy efficiencies; thus, variations in these three variables account for the fluctuating energy intensities in the region. Per capita income is plotted on the X-axis. By plotting the values that correlate per capita GDP with energy intensity, an energy efficiency path is created over time for each country.

Two additional points need to be made here. First, in past studies, similar frameworks have used the purchasing power parity (PPP) measure of GDP instead of using exchange rates. Initially, this study used both PPP rates and exchange rates to plot EEPs. However, since both measures demonstrated similar trends, we opted to use the nominal exchange rate measure of GDP (in constant 1990 dollars) for all calculations (due to better data availability). Second, the total primary energy supply (TPES) is made up of indigenous production, plus imports, minus exports, plus international marine bunkers and changes in stocks. We include combustible renewables and waste in the calculation of TPES for several reasons. Firstly, combustible renewables account for a large percentage of the total energy used in this region. Moreover, development decreases the use of combustible renewables. This results in increases in commercial energy usage, which then leads to economic growth. Hence, if combustible renewables were not in the TPES calculation we would see sudden sharp increases in energy use due to the relationships described earlier. However, by including renewables, the transition to commercial energy is not accounted for.

Findings

Figure 7.1 presents the results of the EEP analysis for five SAARC countries — Bangladesh, Nepal, India, Pakistan and Sri Lanka — over a 27-year period from 1971 to 1997. Maldives and Bhutan were not included due to data limitations. Figure 7.2, found later in the chapter, presents a composite global picture for the year 1997, where the EEPs of five other countries from different regions across the world are compared with that of the SAARC countries.

Figure 7.1
Energy Efficiency Paths (EEPs) for selected SAARC countries

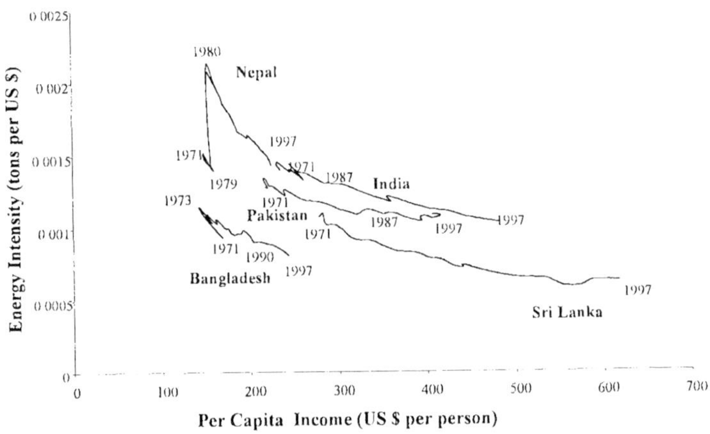

Data Source: IEA (1999)

The optimum EEP should proceed towards higher per capita incomes and low energy intensities (bottom right corner). A number of factors influence trends in energy intensities, most of which directly or indirectly relate to economic growth, or lack of it. Energy intensities decline as a result of structural changes in the economy, such as the transition from energy-intensive manufacturing to less energy-intensive service-oriented economies. Changes in the energy mix from less efficient fuels (coal) to more efficient fuels (natural gas) and improvements in the efficiency of energy production and end use technologies also decrease energy intensities. Finally, policies that promote efficient energy use, such as increased fuel prices, vehicle emissions standards, and renewable energy subsidies, may also serve to decrease energy intensity (Goldemberg and Reid, 1999). Thus, declining energy intensity could either be a component of or a stimulus toward economic growth. The positive ties between economic growth and lower energy intensities are further exemplified in Figure 7.2, where the countries with advanced economies display low energy intensities.

Sri Lanka: From 1971 to 1997 Sri Lanka had the best-performing energy efficiency path in South Asia. Sri Lanka has progressed to achieve the highest per capita incomes and lowest energy intensity levels in the region. This is remarkable, considering that the nation's development effort has been strained by an ongoing separatist's war which has caused almost 60,000 deaths in the last twenty years. Sri Lanka's growth is even more intriguing given that the EEPs of all SAARC countries, including Sri Lanka, were at generally similar levels as recently as 1971. How, then, has Sri Lanka been able to out-perform the rest of the region?

One reason might be its early move toward a market-based economy. In 1977, Sri Lanka was the first SAARC country to make a transition from an inward-looking socialist system to a market economy based on liberalized trade, exchange arrangements and a deregulated foreign investment regime (www.worldbank.org; www.adb.org). In the energy sector, with no proven fossil fuel reserves, Sri Lanka utilized its reservoirs, steep drops and waterfalls to build hydroelectric dams starting in the late 1970s. Today, hydroelectricity contributes 90 percent of Sri Lanka's total electricity generation (CEB, 1998). Though this technology is not entirely environmentally benign due to local environmental and social causes (McCully, 1998; Roy, 1999; Kotagama, this volume), the policy has helped Sri Lanka improve its energy efficiencies and save valuable foreign exchange.

Sri Lanka's growth may in some way be attributed to the so-called 'small island mentality' that has been used to explain the spectacular growth achieved by Japan and Singapore. Being a 'small island' that is densely populated with 18 million people, with limited or no natural resources, Sri Lanka has no choice but to improve its overall system efficiencies in order to sustain growth and remain competitive regionally and globally.

Nepal: In contrast, Nepal has the worst EEP in the region. One of the reasons for this lag in economic and energy efficiency could be traced back to British colonization. Never directly colonized but exploited by the colonists nonetheless, Nepal had to deal with the drawbacks of colonization (for example, the exploitation of its natural resources), without profiting from its few benefits, such as infrastructure development. Thus, Nepal emerged from a state of complete isolation in the early 1950s, when it had among the worst initial conditions for development in the region (www.adb.org). In the early 1980s, Nepal tried to accelerate economic growth through expansionary fiscal policies, but this led to considerable macroeconomic instability (www.worldbank.org). Furthermore, political turmoil brought reforms to a halt. Since then, the rapid political turnover and lingering instability have created both a crisis of continuity and a crisis of neglect in Nepal's energy planning system.

Nepal's low EEP can also be attributed to the structure of its economy. Agriculture and industry account for more than 63 percent of its GDP (UNDP, 1999); this gives it the largest combined percentage in the region (with exception of Bhutan, which earns 74 percent of its GDP from agriculture and industry). Nepal is thus straddled with an economic structure based on relatively energy-intensive activities. Only one percent of households were served by electricity in 1956, and in the last 45 years this has only increased by 14 percentage points (World Bank, 1998). The major energy sources in Nepal continue to be non-commercial: fuelwood, agriculture waste and livestock dung contribute to approximately 90 percent of its total energy share (IEA, 1999). Use of this low-efficiency energy source increases energy intensity, seriously restrains economic development, leads to environmental degradation, causes health problems (especially to children and women), and is not the most environmentally-sound use of these resources. Considering the vast hydroelectric power potential from abundant and largely untapped water resources, Nepal's continued heavy reliance on combustible renewables is both odd and unfortunate (Dixit and Gyawali, this volume).

As Figure 7.1 shows, Nepal exhibits an extremely steep downward sloping EEP curve, indicating the fastest decline of energy intensity in the region. 'Technological leapfrogging,' or the 'follower advantage' are the terms economists might use to explain this steep EEP slope (Goldemberg and Reid, 1999). For example, Nepal is still using combustible renewables, but with better technologies (such as highly efficient stoves) than used in earlier years by the rest of the region. Furthermore, because Nepal is 'very dirty' (as in highly energy inefficient), it is easier and cheaper to adapt new technologies. However, Nepal's EEP slope will gradually decrease when it becomes harder and more expensive to adapt new technologies. Despite this steep downward EEP curve, Nepal's 1997 EEP level is only comparable to India's 1971 levels. Thus, if we assume that EEP is an indicator of development, then Nepal is roughly 27 years behind India and 43 years behind Sri Lanka in achieving the level of energy security that these two countries have attained.

Bangladesh: Bangladesh demonstrates a unique EEP in comparison to the rest of the region. Its energy intensity levels are similar to those of Sri Lanka (the highest in the sample), while its per capita income levels parallel Nepal (the lowest in the sample). It seems strange that a country with a highly efficient energy intensity path should also have one of the lowest per capita income paths in the region.

Economic problems faced by Bangladesh are mostly structural. In the 1990s it tried to establish a more open, market-based economy. Unfortunately, despite these reforms overall business confidence remained fragile due to political disturbances, a slackening in structural reforms, poor governance, and a generally slow pace of reform — especially in areas of privatization, the financial sector, and the trade regime (www.worldbank.org; www.adb.org). Although these reforms did not bring about the expected results, the Bangladeshi EEP has significantly improved since 1990. Bangladesh's low energy intensity levels are attributed to its energy mix. Natural gas, which has high efficiency levels, accounts for almost 25 percent of the total primary energy share. With recent energy reforms that encourage the increased use of natural gas as a substitute for imported oil, Bangladesh is likely to see energy intensity levels drop even further.

India and Pakistan: Indian and Pakistani EEPs were at relatively similar levels in 1971 (Pakistan's energy intensity level was slightly lower than India's); by 1997, however, the Indian EEP had outperformed the Pakistani EEP. This is remarkable, considering that India is one of the developing world's largest consumers of energy and is among the top 20 users of commercial energy in the world (EIA, 1998). India relies on coal, a highly inefficient fuel, for 55 percent of its total primary energy supply. On the other hand, Pakistan relies on natural gas, a highly efficient energy source, for 25 percent of its total primary energy supply (EIA, 1998). With such an energy mix, Pakistan's EEP should be far less energy-intensive than India's.

The expectation that Pakistan's EEP should have performed better than India's held true up to 1987. Since then, however, Pakistan has seen stagnation in energy intensity levels and dismal growth in per capita incomes, while India has improved its performance on both counts. Pakistan's stagnation period from 1987 to 1997 coincides with massive political instability. During this period there were six changes of government and the political, social and economic systems were in chaos. Thus, one of Pakistan's biggest problems has been its now chronic internal political instability.

Global Comparisons: In Figure 7.2, South Asian EEP levels for 1997 are compared with five selected countries from around the world — the US, Sweden, Argentina, China and Nigeria — to provide context to the discussion. These countries were chosen from five continents with varying levels of development. Clearly, there are strong correlations between the energy efficiency and the economy: stronger economies invariably have better energy efficiency levels. For example, the US and Sweden, two advanced economies, have low energy intensity levels (high energy efficiency levels). Argentina shows an EEP level that is progressing toward that of an advanced economy. China is on par with most of South Asia, which has a relatively poor EEP level compared to the rest of the world. Nigeria, one of the 20 poorest countries of the world, has the highest EEP level (low energy efficiency levels). Since it is an oil-rich country, Nigeria may have a tendency to overuse its energy, thereby increasing its energy intensity. In addition, political uncertainty, poor governance, widespread corruption, and inefficient state-owned firms in the oil, electricity and telecommunications sectors have also contributed to Nigeria's poor economic performance and low energy efficiency levels.

Figure 7.2
Energy efficiency levels for selected countries (1997)

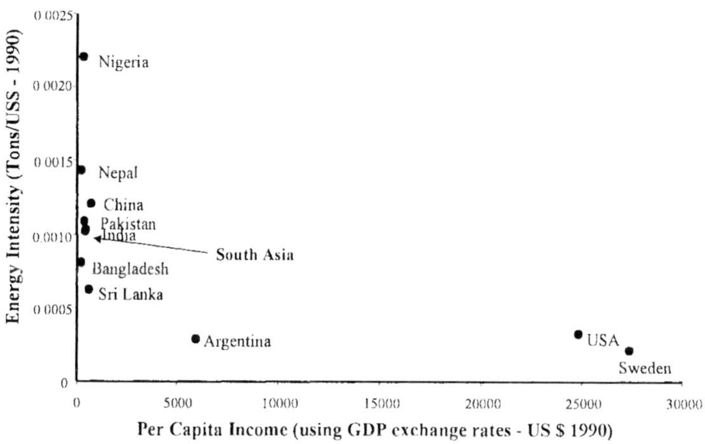

Data Source: IEA (1999)

Summary: The energy efficiency path analysis provides one perspective on understanding why different SAARC nations are at varying levels of energy security. Though it does not provide us with perfect, clear-cut answers, it underscores several key trends in energy security. There are many reasons for the regional differences in energy efficiency paths, including technological improvements in individual energy end-use and conversion components, such as efficient stoves (in the case of Nepal) and/or more efficient energy generation. They also include structural shifts in energy systems; for example, shifts from coal to gas. Other factors that have influenced the EEPs include: inter-fuel substitution at the level of energy end use, such as the conversion from the use of biomass to gas or electricity, and structural shifts in the economy as in movement toward less energy-intensive activities, which ultimately changes the patterns of energy end uses and lifestyles. Although not a conclusion in the scientific sense, this analysis suggests strong correlations with issues of good governance. Political stability, policy continuity and corruption are certainly important factors.

The analysis also suggests a correlation between economic development and energy use efficiency. Unfortunately, poorer countries also tend to be less efficient in converting energy into wealth. One might postulate that the same sad relationship can be transferred to the poorest communities. On the other hand, South Asia could be content by the fact that all countries display downward-sloping, progressive energy efficiency paths, which demonstrate a push towards a better use of energy in terms of contribution to economic growth.

Energy Dependence Paths (EDPs)

In this section we analyze energy and security in South Asia in terms of energy dependence. Energy dependence paths are calculated by plotting each country's energy imports per GDP as a function of per capita international liquidity. Countries with a lower level of dependence on imported energy and higher reserves of international liquidity are considered to be comparatively more 'energy secure'.

Energy Dependence as Energy Insecurity

Energy security is typically defined in terms of countries possessing, or having the ability to acquire from domestic or foreign sources, adequate energy reserves and production and distribution facilities in the foreseeable future (Belgrave, et al., 1987). Insecurity arises when energy users' day-to-day activities are threatened by supply shortages or sudden and major price variations. The term 'energy security' first gained publicity during the energy crisis of the 1970s, which severely affected both industrialized and developing economies. For South Asia, global energy crises can adversely affect four kinds of basic security concerns — economic, strategic and military, social, and political security (Thomas and Ramberg, 1990).

Economic Concerns: Economic security may be defined from an energy dependence perspective as national resource sufficiency and accessibility to goods and services in world markets at affordable prices. More specifically, the question is whether a nation has sufficient economic endowments to 'pay' its energy bill without compromising other national priorities, particularly developmental ones. For the SAARC nations, periodic spurts in OPEC oil prices produced extreme imbalances in foreign trade and severe dislocations in domestic economic plans. In most South Asian countries the oil import bill rose from 20-30 percent all the way to 50-80 percent of their exports earnings during the late 1970s (Thomas, 1990). For example, in Pakistan the oil imports bill skyrocketed from 50 million US dollars in 1973 to 540 million dollars in 1979, and again to 1.1 billion in 1979-80 following the Iran crisis (Ebinger, 1990).

Political Concerns: Political security is defined as the maintenance of domestic stability, either by an elected or authoritarian government (Thomas, 1990). While absolute evidence is difficult to conceive, it is noteworthy that the economic crisis resulting from the energy crisis correlated with periods of political instability and insecurity in the region during the 1970s. While many other forces also came into play, the diversion of significant portions of national foreign exchange reserves to energy imports contributed heavily to the economic hardships which, in turn, added further stress on already fragile economies and polities.

The point, however, is that sudden diversions of large amounts of foreign exchange to energy purchases can have significant impacts on political stability at home for fragile developing economies. For example, the anecdotal evidence suggests that in 1975 and 1977 economic stresses and public dissatisfaction led to the declaration of 'emergency' status in India. In Sri Lanka, Janatha Vimukthi Peramuna (JVP), a social movement dedicated to the liberation of Sri Lanka from the worsening socio-economic and political crises, attempted several social revolts (Gunaratna, 1995). Similarly, the regime in Pakistan began to feel the pressure of political liberalization and potential revolution as a result of the economic crisis. While these may or may not have been linked to global energy costs, they were certainly linked to widespread feelings of economic depravation.

Social Concerns: Social security may be viewed as a function of the resilience of a community to adapt to external perturbations. Resilience is a measure of the persistence of systems and their ability to absorb change and disturbance while maintaining the same relationships between population and state variables (Hollings, 1973; Pimm, 1984; Perrings and Opschoor, 1994). Global energy crises like the one in the 1970s affected societal resilience in many South Asian countries. To increase depleted foreign exchange reserves, these countries encouraged unskilled and semi-skilled laborers to work in the oil-producing countries of the Middle East. Most jobs in the Middle East called for domestic assistance; hence, many young South Asian women (especially from Sri Lanka and Bangladesh) migrated, leaving their families behind. As a result the family structure was severely weakened and in turn led to severe societal problems, including, physical, sexual, and emotional abuse of children by their fathers as well as increased levels of divorce and suicide.

Strategic and Military Concerns: The degree and intensity of perceived threats and the ability of the military to meet these challenges are used to gauge the strategic and military security of a nation. When faced with increasing external threats, a nation's domestic resources are diverted away from necessary development programs in order to counter these threats. During the 1970s energy crisis, the Western and Eastern blocs increased arms exports to reduce their trade deficits. Concurrently, oil-exporting countries such as Iran, Saudi Arabia, Iraq and Libya began to buy sophisticated arms with their surplus petrodollars. This arms race spilled over to the SAARC region as well (Thomas and Ramberg, 1990).

Mass inflows of armaments into India and Pakistan during this period were a function of their long-standing differences, but were made easier by the then-flourishing arms trade. In turn, this has had a spiraling effect throughout the region, which has become more militarized. Increased military spending has diverted funds from other developmental programs. In 1997, the SAARC countries spent approximately 20 billion US dollars on their militaries, while an estimated 580 million people in the region are surviving on less than one dollar a day (UNDP, 1999). Once again, while energy is certainly not the prime motor of these events, it was one among many contributing factors and needs to be recognized as such.

The EDP Framework

To better understand the links between energy security and energy dependence in the region, we will utilize an energy dependence path framework. A scatter plot time series model is used to plot energy dependence paths by plotting the total net imports per GDP as a function of per capita international liquidity. Figure 7.3 plots EDPs for five SAARC countries — Bangladesh, Nepal, India, Pakistan and Sri Lanka — over a 27-year period from 1971 to 1997. Maldives and Bhutan were not included due to data limitations. The Y-axis plots the ratio of total net energy imports (I) of carbon equivalent, in tons, to gross domestic product (GDP) in US dollars, or I/GDP. An exchange rate measure of GDP in constant 1990 US dollars is used.

Figure 7.3
Energy Dependence Paths (EDPs) for selected SAARC countries

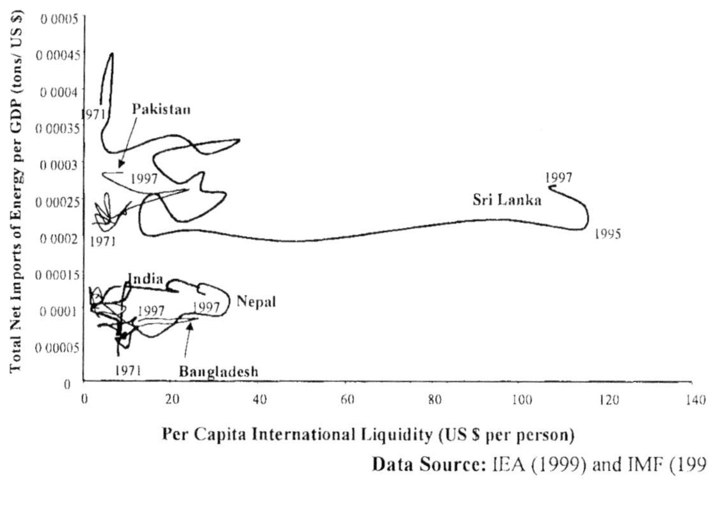

Data Source: IEA (1999) and IMF (1999)

A lower value of I/GDP indicates less dependence on imported energy. Per capita international liquidity is the ratio of total reserves minus gold (R), in US dollars, to population (P), or R/P (US dollars per person); this is plotted on the X-axis. By connecting the values that correlate net energy imports per GDP with per capita liquidity, an energy dependence path is plotted for each country. International liquidity or hard currency is important for these countries, as the purchase of imported energy requires adequate reserves of hard currency. Consequently, less dependence on imported energy and high levels of international liquidity lead to higher energy security levels.

Findings

In deciphering Figure 7.3, note that low levels of total GDP dollars per ton of net imported energy (I/GDP) indicates reduced vulnerability to external fluctuations in the global energy markets. Furthermore, most imported energy is purchased using hard currency, as measured by international liquidity per person. Thus, high levels of per capita international liquidity and low levels of I/GDP are vital for the sustained energy security of a nation. The SAARC region accounts for approximately 20 percent of the world's population but only 4.9 percent of global primary energy consumption (EIA, 1998). However, rising populations (expected to double by 2050), vigorous economic growth (5.5 percent per year, on average) and steady shifts from non-commercial to commercial energy sources are all expected to rapidly increase the demand for commercial energy in the region.

Pakistan: In Figure 7.3, Pakistan demonstrates the weakest energy dependence path; i.e., it is the most energy dependent. In 1997, Pakistan was highly dependent on imported energy, despite possessing vast reserves of natural gas. In addition to this dependence on imported energy, Pakistan's per capita hard currency reserves are one of the lowest in the region. Pakistan's heavy dependence on imported energy can be attributed to two factors: its geography and the mismanagement of its energy sector. Geographically, Pakistan has a large hydroelectric potential, with estimates varying between 10,000 and 12,000 megawatts (Ebinger, 1990). However, hydropower development has proved troublesome for Pakistan due to large seasonal fluctuations in river flows; droughts (especially since irrigation gets priority over power generation); and high siltation rates, which limit the life of hydro plants (Ebinger, 1990). Furthermore, Pakistan's significant coal reserves are all poor quality lignite or sub-bituminous coal with limited coking potential, low calorific content, and a high sulfur component. Given these dilemmas, Pakistan has opted to develop a substantial thermal electricity generating capacity, which has only increased its dependence on imported oil.

Though Pakistan has the highest per capita production of natural gas in the region, it has not optimized the use of its abundant natural gas reserves, is particularly wasteful of it, and has not utilized the resource efficiently. Ironically, Pakistan is currently facing an oversupply of power due to poor energy sector planning. A number of independent power projects (IPPs) are expected to produce more than 3000 MW of additional power in the years ahead. Contractual obligations to buy electricity from IPPs will further burden the weakened economy. Thus, mismanagement in the energy sector has rendered Pakistan more vulnerable to changes in the global energy markets in 1997 than it was in 1971. Its dependence on imported energy has increased significantly during this period, while the reserves of hard currency used to purchase this energy have barely changed between 1971 and 1997. In effect, Pakistan was using far more of its hard currency reserves on energy in 1997 than in 1971.

Sri Lanka: Sri Lanka shows a similar energy dependence profile as Pakistan, but demonstrates considerable improvement in its EDP from 1971 to 1997. Sri Lanka depends on imported fossil fuels, as it has no proven reserves of its own. However, Sri Lanka's primary source of commercial energy is hydropower. The addition of new hydropower plants has resulted in sharp decreases in Sri Lanka's dependence on imported energy. The sharp improvement of its EDP in 1976 was due to the commissioning of two new hydropower plants — New Laxapana and Ukuwela — with a combined generation capacity of 650 gigawatt-hours per year (GWh/year). Similarly, sharp improvements in Sri Lanka's EDP curve in 1986 and 1989 coincide with the opening of five other new hydropower plants with an estimated combined capacity of 1800 GWh/year (CEB, 1998).

However, Sri Lanka's reliance on hydropower makes the country highly vulnerable to rainfall fluctuations. Failure of monsoon rains leads to decreased hydropower generation, which then increases the demand for thermal energy. For example, low rainfall in 1987 and 1995 led to sharp increases in energy dependence as seen in the EDP trajectory. From 1986 to 1987 Sri Lanka's hydro generation dropped from 99.7 percent of the total electricity generation to 80.4 percent, while thermal generation increased from 0.3 percent to 19.6 percent. Similarly, from 1995 to 1996 hydro generation dropped from 94 percent of the total to 71.8 percent, while thermal generation increased from 5.6 percent to 24.9 percent (CEB, 1998). Moreover, most of Sri Lanka's hydropower potential has already been harnessed; further exploration of hydro resources may be increasingly difficult due to impacts on the environment and ecosystem (CEB, 1998). Hence, one would only expect Sri Lanka's energy dependence to increase, rather than decrease, with time.

Bangladesh: In the early 1970s the EDPs of Bangladesh, India, and Nepal all displayed similar trajectories; however, their paths have diverged over time. In 1997, Bangladesh (with Pakistan) had the lowest hard currency levels in the region. Yet unlike Pakistan, Bangladesh's dependence on imported energy was quite low. This low level of dependence on imported energy is attributed to two factors.

On the one hand, nearly 65 percent of primary energy in Bangladesh is supplied by biomass fuels and only 35 percent by commercial sources. On the other hand, and perhaps more importantly, of the primary energy supplied by commercial sources, only 10 percent comes from imported oil (EIA, 1998), while 25 percent comes from natural gas; Bangladesh is the second highest per capita producer of natural gas in the region. The impact of hard currency reserves on the Bangladeshi EDP mirrors its economic growth patterns, as in the case of Sri Lanka. From 1990 to 1994 the trends have been moving in the right direction. The fiscal deficit fell from 7.7 percent to 6 percent, inflation remained moderate within the 1.3 to 5.2 percent range, the current account deficit dropped from 6.9 percent to 1.6 percent of GDP, and most importantly, foreign exchange reserves rose from 1.9 to 2.8 months' worth in imports (www.worldbank.org). However, prolonged political disturbances during 1995-96 had a negative impact on macroeconomic performance; these patterns of economic growth correlate well with shifts in the Bangladeshi EDP curve during the periods mentioned above (1990-1994, 1994-1997).

Nepal: Nepalese dependence on imported energy is relatively low, since biomass supplies 90 percent of the total primary energy there. The upward trend of Nepal's EDP up to 1993 is partly due to structural economic transformation. In 1997, hydropower accounted for only 1.3 percent of the total primary energy supply, while oil accounted for 8 percent (EIA, 1998). Given Nepal's relatively small export base, roughly a quarter of the foreign exchange earned from exports is required to cover costs of imported oil (www.adb.org). The sharp reversal of Nepal's EDP after 1993 is related to the political atmosphere at that time. Political turmoil in 1994 and 1995 blocked reforms that attempted to establish a market-based economy. The weakening fiscal situation, poor agricultural performance, and a sharp decline in exports all contributed to the erosion of the economy. This also reduced hard currency reserves after 1993, as seen in Figure 7.3.

India: The Indian economy has been growing steadily since 1991, when economic liberalization began. Reforms and heavy monsoons increased GDP growth to 5 percent in 1992-94 and to 7 percent in 1994-96, placing India among the world's best performing economies (www.worldbank.org). The steady growth in the Indian economy is indicated by its EDP trajectories, which show a steady progression since 1991. As for India's energy mix, coal and oil account for most of its commercial energy generation. India has substantial coal and oil reserves, but production has been relatively stagnant since 1995, rendering the country more dependent on imported oil. Similar to Nepal, about a quarter of its foreign exchange earnings is used to buy oil (WRI, 2000).

Global: To place the energy dependence paths in the SAARC region into perspective, Figure 7.4 plots the EDPs of four other countries — the US, Argentina, Sweden, and China — in comparison to the SAARC region in a composite global graph for the year 1997. Argentina (which has vast oil reserves) and Sweden (which is highly energy efficient) are the most energy-secure countries. Correspondingly, Pakistan and Sri

Lanka are the least energy-secure. China is progressing towards a better EDP. The rest of the South Asian countries have comparatively high levels of imported energy dependence with low hard currency reserves with which to purchase this energy. In essence, South Asia remains a significantly energy-insecure region in comparison to others; it is likely to become even more so in the future.

Figure 7.4
Energy dependence levels for selected countries

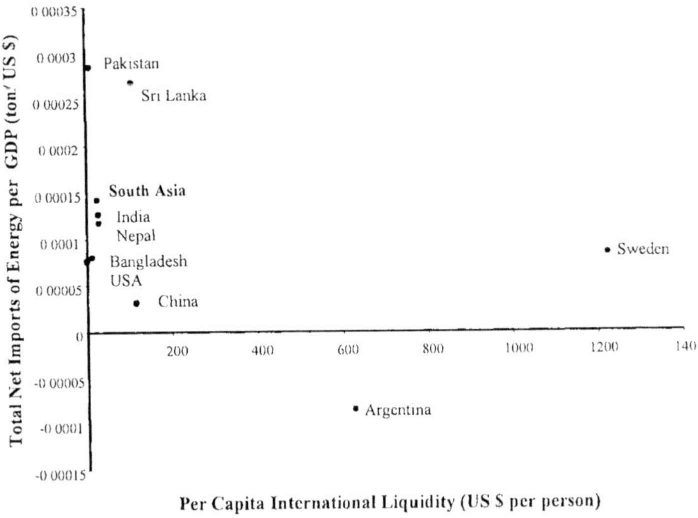

Data Source: IEA (1999) and IMF (1999)

Summary: South Asia suffers from a high level of energy insecurity in terms of its dependence on outside sources for large proportions of its energy needs. This is particularly troublesome given the poverty of the region and the likelihood of it to become even more dependent in the future. Among the options available to the SAARC countries in this regard, the most important one might be the choice of energy mix. It is important to develop domestic commercial energy sources and to utilize them prudently. Unfortunately, the evidence has not been very encouraging before now — for example, Pakistan's use of natural gas, Nepal's use of hydroelectric and India's use of coal have all been suboptimal. Figures 7.3 and 7.4 highlight the region's high vulnerability to global energy markets. Indeed, the challenges of energy dependence may be the most significant to the region's energy security.

Energy Impact Paths (EIPs)

The final framework we use to analyze the energy security situation in South Asia relates to the environmental impacts of energy by charting the energy impact paths of five SAARC countries. In looking at energy security through this lens we hope to uncover ideas and issues that might not have been properly dealt with in the earlier frameworks, which focused on economic efficiency concerns and external dependence issues, respectively. In looking at the energy security problematique from the standpoint of environmental impacts, we follow the lead of a number of earlier chapters by taking a decidedly human security view of things and placing it within the context of sustainable development.

Energy as the Motor of Sustainable Development

Discussions linking environment and development concerns took the global center stage after the publication of the Brundtland Commission's Report (WECD, 1987) which articulated the ideal of 'sustainable development' and defined it as "development that meets the needs of the present without compromising the ability of future generations to meet their own needs." In fact, the concept of 'sustainable development' paved the way for the 1992 United Nations Conference on Environment and Development (UNCED) in Rio de Janeiro, which was the first meeting where issues of environment and development were simultaneously addressed (Banuri, 1992; Thompson, 1992; Adede, 1992; Najam, 1993). Consequently, environmental issues now play an important role in all aspects of developmental decision-making and visa versa. For example, most development projects undergo environmental assessments; the most environmentally sensitive approach is often favored over less sensitive ones. Furthermore, social issues such as poverty, which were once considered purely developmental, are now also recognized as environmental issues. Environmental concerns such as climate change are now also seen as developmental issues. Thus, notions such as 'pollution of affluence' (by way of affluent lifestyles) and 'pollution of poverty' (due to lack of development) have gained global acceptance (Munasinghe, 1993b; Dasgupta and Maler, 1990; World Bank, 1992).

At the same time, environmental concerns about energy production and consumption have shifted from regional, short-term problems (air pollution) to long-term, global concerns such as climate change (Anderson, 1994; Clark, 1991). At the same time increases in energy use remain closely related to a country's economic development. Thus, to a certain extent economic development is a 'double-edged sword' which can be both a solution to environmental problems stemming from poverty and a cause for new environmental problems such as climate change (Munasinghe, 1993a; Strong, 1992). Accordingly, the crucial dilemma facing the developing world, including South Asia, is to reconcile the developmental paths they have chosen with responsible environmental stewardship (Munasinghe, 1999a).

The EIP Framework

An energy impact path framework is used in this section to understand the energy impacts on sustainable development in South Asia. The framework is modeled after the Environmental Kuznets Curve (EKC) hypothesis, which proposes an inverted U-shaped relationship between environmental degradation and income per capita (World Bank, 1992). To differentiate between energy impact within the region, Figure 7.6 plots per capita carbon dioxide emissions as a function of per capita income for the seven SAARC countries — India (1971-96), Pakistan (1972-96), Sri Lanka (1971-96), Bangladesh (1972-96), Nepal (1971-96), Maldives (1977-96) and Bhutan (1980-96).

The Y-axis is defined as the ratio of total carbon dioxide emissions (from fossil fuels, cement manufacture, and gas flaring) (C) in metric tons of carbon to population (P) in number of persons, or C/P. Increases in C/P are primarily attributed to increased energy use resulting from economic development. The x-axis plots per capita income, which is used as a measure of wealth, reflecting economic development. Figure

7.6, therefore, gives us an energy impact path for each country in the sample.

The original Kuznets curve hypothesis suggested that a measure of inequality in the distribution of income is related to the income growth by an inverted U-shaped curve (Kuznets, 1955). The idea was that inequity would increase as countries developed, but at some threshold further development would decrease societal inequity rather than increase it. Today some economists have expanded on this hypothesis and propose an inverted U-shaped relationship between environmental quality (or pollutant emissions) and economic growth — this notion is often referred to as the Environmental Kuznets Curve hypothesis (Panayotou, 1993; World Bank, 1992).

Figure 7.5
Environmental Kuznets Curve

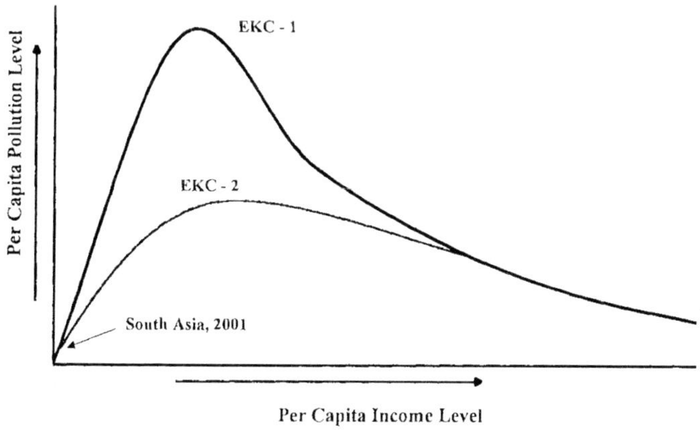

Figure 7.5 tries to notionally illustrate the EKC hypothesis. The idea is that a country at low levels of economic activity is likely to have low environmental degradation, which is then likely to increase alongside economic development. It will finally decline at some threshold as the fruits of new prosperity are invested into environmental cleanup. Proponents of the idea present two explanations to justify the inverted U-shape of the curve. The first is based on the assumption that the environment is a 'luxury good', and in the early stages of development

people are unwilling to trade consumption for investment in the environment. However, once people reach a level of consumption known as the 'income turning point', they become increasingly proactive and wish to improve the quality of the environment. The second explanation for the shape of the curve relates to stages of economic growth. This refers to the structure of the economy as it progresses from agricultural to industrial, and finally post-industrial service-based economies. A transition from agriculture to industry increases environmental degradation, while a transition from industry to the service sector reduces degradation. A similar view applies the transition of economies through 'technological life cycles' such as a progression from smokestack technology to internet technology .

The EKC hypothesis was popularized by the World Bank in its 1992 *World Development Report*, but has attracted a growing number of critics over the last decade. Stern (1996) critiques the weakness of the empirical evidence and inappropriateness of the statistical techniques used; Moomaw and Unruh (1997) argue that the income/environment relationship in the EKC is merely a result of polynomial curve fitting (Arrow, et al., 1995; Pearson, 1994; Pezzey, 1989).

Despite increased criticisms of the conceptual and methodological limits of EKC analysis (which the authors share) we use it nonetheless in this study. Unlike its proponents, we do not believe that the inverted U-curve has predictive value. Instead, we believe that different countries at different points in time can, and do, adopt different routes to transition and, indeed, that progressive and forward-looking policies can allow nations to 'tunnel' through the EKC. In essence, we utilize the concept not because we believe it has predictive capacity but because the relationship it posits is important to understand.

Two points need to be highlighted here. First, if environment and income links can serve as another expression for 'stages of economic growth,' then comparative EIPs will enable us to differentiate between developmental levels while analyzing the sustainability of respective development paths in the SAARC region. Second, and more importantly, South Asia is at the very early stages of development and is primed for rapid economic growth. This projected economic growth will invariably increase the energy consumption potential of its 1.4

billion people which in turn will burden the environment immensely. Hence, following the expected boom, South Asian economic development could take either the unsustainable EKC-1 path or the relatively more sustainable EKC-2 path (Figure 7.5), or conceivably some even more sustainable path to 'tunnel through' EKC-1. Thus, the emphasis should be on identifying policies that would help de-link environmental degradation from growth so that the environmental harm would be reduced along the developmental path (Munasinghe, 1995b; 1999b). By analyzing individual EIPs within the SAARC region, the sustainability of various developmental paths can be identified.

Figure 7.6
Energy Impact Paths (EIPs) for SAARC Countries

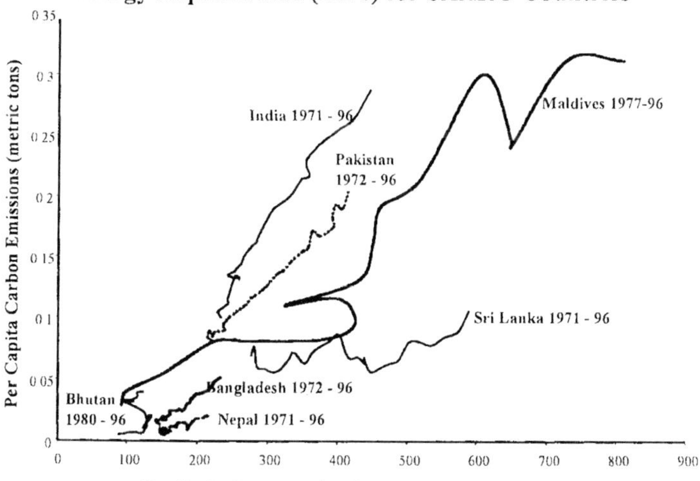

Data Source: IEA (1999) and IMF (1999)

Findings

Interestingly, it is evident from Figure 7.6 that there are three distinct developmental levels within the SAARC region. Bhutan, Nepal, and Bangladesh are at the lowest levels of economic development. These countries demonstrate relatively low levels of pollution as well as low per capita incomes. India and Pakistan are at middle levels of economic development, with per capita incomes in the center ranges. Sri Lanka and Maldives have the highest per capita incomes in the region; they correspondingly fall into the highest category of economic development. More importantly, two levels of sustainability in EIPs exist within the region. Figure 7.6 suggests that the EIPs of India and Pakistan, for example, exhibit slopes that conform to a relatively steeper (i.e., more unsustainable) path to development, resembling EKC-1 in Figure 7.5. However, the EIPs of Sri Lanka, in particular, shows a slope that corresponds to a relatively more sustainable path, akin to EKC-2 in Figure 7.5.

Sri Lanka's apparent ability to considerably improve its per capita income levels (1971 to 1996) while maintaining stable pollutant levels is quite impressive. Essentially, Sri Lanka has managed to rapidly develop its economy without unduly burdening the environment. Since CO_2 emissions are related to energy production and consumption, Sri Lanka has made a more efficient use of its energy for economic development. In contrast, India and Pakistan have put relatively more stress on the environment despite low levels of economic development. Differential sustainability levels could be attributed to a number of reasons, including the structure of the economy, the energy mix and levels of efficiency.

Figure 7.6 suggests that India and Sri Lanka follow divergent sustainability paths. Essentially, India's EIP slope is much steeper than that of Sri Lanka. Identifying factors that contribute to the carbon dioxide emission paths of India and Sri Lanka may help determine reasons underlying the disparity. Bangladesh and Nepal, who are at relatively lower levels of development, have EIPs that seem prepared to follow a relatively more sustainable path, but only if future policies and institutions are designed specifically to achieve such results.

One should note that in comparison to the advanced industrial countries, all countries of South Asia are quite sustainable in their impacts on the global environment. All are still early in their developmental 'climb' and none are major polluters of the global commons. The challenge is to resist the trend followed by today's industrialized economies and seek new paths to development whose impacts are more benign to the environment. The goal for the region should be to 'buck' the EKC trend, if there is one, and to 'tunnel through' it via more sustainable development paths that allow countries to meet the developmental aspirations of their current and future generations without sacrificing environmental quality. "With such a focus, the EKC analysis becomes useful mainly as a metaphor or framework for policy analysis, while other issues become less important – such as the exact shape of the EKC, or whether the empirically estimated EKCs which tend to be based on a cross section or pooled data can adequately capture the growth characteristics of any single country" (Munasinghe, 1999b).

Looking Forward

In this paper we have focused on several questions pertaining to trends in South Asian energy security. Specifically, we have been interested in differentiating between the various energy security paths observed in the region in addition to examining the causes for these differences. In doing so, we analyzed the energy *efficiency* paths (EEPs), energy *dependence* paths (EDPs), and energy *impact* paths (EIPs) for different SAARC nations and for the region.

While the conclusions of each individual framework have already been highlighted in the sections above, our findings suggest that there are a number of key ingredients to a progressive energy policy that is likely to make developing countries, such as those in South Asia, move toward the twin goals of sustainable development and enhanced energy security. While these ingredients are tentative in some respects, they are supported by the larger literature as well as by the specific findings of other chapters in this volume.

Energy

Structural shifts in the energy sector. A shift toward indigenous energy sources (hydropower in Nepal and Bhutan and natural gas in Pakistan) and energy-efficient commercial energy sources (e.g. a shift from coal to natural gas in India) will improve energy security and energy intensity levels.

Technological improvements. While there are no easy 'technofixes', prudent technological improvements in energy end use (e.g. more efficient stoves and lighting) and conversion components (power plants) will enhance energy efficiencies.

Promote inter-fuel substitutions. At the end-use level (e.g., from biomass to gas or electricity) inter-fuel substitution can increase efficiency and put resources to more appropriate use while promoting economic growth.

Economy

Structural shifts in the economy. A shift toward less energy-intensive but more capital-intensive activities, for example a shift towards the service sector, is likely to decrease energy usage while increasing economic growth; economic liberalization, when conducted with sensitivity to its social impacts, also serves to improve economic efficiency and free up more resources for development purposes.

Positive balance of trade. This can be achieved through increased exports (of goods and services) that would increase the reserves of 'hard currency', which in turn leads to enhanced buying power (of imported energy). Thus, a country with a positive balance of trade becomes less vulnerable to global energy market fluctuations.

Governance

Stable political atmospheres. Our study strongly suggests that political stability, or lack thereof, can have direct and drastic impacts on energy security paths.

Integrated planning. Piecemeal and ad hoc energy planning has taken a high toll in South Asia and sometimes leads to contradictory policies in related fields that not only cancel out each other's effects, but can be even more detrimental in combination. There is a clear need for integrated economic, energy and environmental planning.

Final Thoughts

Perhaps the most important conclusion to be drawn from our study is that a sustainable approach to the inevitable increases in energy consumption is possible within South Asia. Sri Lanka's success in sustainable energy consumption should serve as an example. Moreover, conditions are still favorable in the region as a whole to make a transition to relatively more sustainable energy paths. This is because all the countries of the region are still in the early stages of development and are, therefore, better able to change their development and energy trajectories.

Finally, it is appropriate to end on a note of caution regarding any of our findings. In adopting a multiple frameworks analysis, we explicitly sought to get a more complete picture of the energy security challenges faced by the region. However, this does not mean that we claim to have presented a complete picture. While we would profess that using multiple frameworks has provided us with a more expanded and multidimensional understanding of the issues, we do not claim comprehensiveness. Indeed, a number of factors that we consider to be critical are not covered in the frameworks we have used. These include, among others, issues of equity, social justice and governance. For example, a major reason for Sri Lanka's high rating in our framework has been its reliance on hydroelectric power. However, as we alluded and as a number of prior chapters have detailed, large dams and infrastructure projects can have serious social ramifications and can be particularly unfriendly to the interests of the poor.

While we recognize that these and other related concerns could not be factored into the frameworks we used, we do believe that analysis using multiple frameworks is superior because it allows for the identification of multiple dimensions of a problem. The challenge remains before us to devise other frameworks that can also accurately incorporate the challenges of equity and social justice.

Chapter Eight

Land-use & Food Security: The Green Revolution & Beyond

KHALID SAEED

FOOD IS A FUNDAMENTAL human need, and the provision of this necessity has been viewed as an essential component of security for individuals and societies. Food security is viewed in this paper as the sustainability of food supply in the face of environmental limitations. Although this definition may not be in dispute, 'food security' appears to have been defined largely as food adequacy in terms of the intents of agricultural policy made by governments. Indeed, agricultural land use practices designed to increase agricultural output have created what is known as a 'green revolution', which has kept food supply at a reasonable level in spite of population growth. Unfortunately, these policies have increased the vulnerability of food supply to external shocks and have even precipitated trends that point toward an impending decline in food production due to the decay of environmental resources. Thus, governmental policy has undermined the food security of the region. Apparently, such policies were implemented without a full understanding of the societal and ecological dynamics that caused the food shortage in the first place. Other developmental policies applied in the region have also been based on a similar perspective, as is amply documented in this book. This has resulted in the decay of human security also in terms of environmental support, economic well-being, health, and law and order. Evidently, threats to human security often arise following the enactment of policies that attempt to alleviate symptoms without addressing the underlying relationships creating the problems at hand.

This chapter revisits the issue of agricultural land use and food security for South Asia by placing it in the context of the Asia-Pacific region as a whole. It analyzes data for a total of fourteen countries, which are categorized as high-income, medium-income, or low-income. The four South Asian countries in the sample -- India, Nepal, Pakistan, and Sri Lanka -- all fall into the third category, which allows us to get a snapshot of the South Asian situation in comparison to the rest of Asia. Reviewing the published time-series data from a larger cross-section of Asia-Pacific countries, an effort is made in this paper to illustrate the pattern of changes taking place in the agricultural resource base of the larger region. The sub-category of low-income countries, which is predominantly assigned to South Asian countries, highlights both the differences and commonalties between the patterns within South Asia as well as in the Asia-Pacific region as a whole. The underlying trends indicate that while the consumption base has expanded across the board through increases in population and income, food production has increased largely through intensive cultivation while agricultural land resources have, for the large part, stagnated. In some instances, the amount of agricultural land used for cultivation has increased, but often at the cost of a reduction in forested land.

The chapter also translates the characteristic relationships underlying agricultural land use policies, and how they affect population growth and agricultural resource systems, into a system dynamics model. Experimentation with this model shows that a sharp decline in agricultural production may be expected across the board, but particularly in the medium- and low-income countries, including those in South Asia, unless sustainable agricultural technologies are adopted. Unfortunately, even when available, such technologies will not be put into practice as long as the technological and economic considerations governing agricultural policy remain divorced from environmental information concerning land resources and soil ecology. If the institutional frameworks and incentives in South Asia remain as they are, it is likely that the short-term private gains in production will continue to be sought at the cost of the decay of the common resource system that sustains agriculture. This chapter concludes, therefore, that new institutions must be formed to connect appropriate ecological information with the economic decisions leading to just and lasting technological choices.

The Food Security Problem

The problem of food security is often perceived in the conventional sense as an existing condition which must be alleviated and which is quite pervasive also in the other developmental contexts. For example, poverty, unequal income distribution, food shortage, high illiteracy, low infrastructure inventory, and corruption are often seen as obvious problems. In a business context, low market share, overstaffing, poor quality, low productivity, non-profitability and the like are all considered problems. In all such cases, the starting point for a quest for new policies is the acceptance of the existing condition, also known as the problem. In a scientific framework, however, a problem is articulated as an internal behavioral tendency found in a system. It may represent a set of patterns, a series of trends, or a set of existing conditions that appear resilient to policy intervention. In other words, an end condition alone cannot be defined as a problem in this sense, rather the problem is the complex pattern of change leading up to this end condition. Developmental problems must therefore be defined in a scientific sense and not assumed to exist if an effective intervention framework is to be achieved. This would require a careful study of past trends, including relevant variables and an intelligent extrapolation of these trends into a future context.

The pattern of trends representing the food security problem articulated in this paper was prepared by the author some time ago for the United Nations Economic and Social Commission for Asia and the Pacific (UNESCAP) (Saeed, 1994). Some 300 studies, covering fourteen selected countries representing Asia and the Pacific region over the past three decades, were constructed from published UN sources to serve as a database for discerning past trends. However, many cells were missing from the data. There were also differences in units of measurement, in definitions of data categories, and variability in national policies. Furthermore, there existed a large variety in the size of the economies and, therefore, data from one country could dominate an aggregate trend. Hence, it was decided not to assimilate national data into regional categories, but rather to closely examine the selected set of countries as many manifestations of the same food security problem.

The selected countries were divided into three categories based on a classification proposed by the Asian Development Bank (ADB) (Okita, 1989). Australia, Japan, Korea, and Singapore were placed in category A, representing relatively high levels of income; Malaysia, Thailand, the Philippines, and Indonesia were placed in category B, representing middle levels of income; and China, Vietnam, India, Nepal, Pakistan, and Sri Lanka were placed in category C, representing relatively low income levels. Category C is largely dominated by members of the South Asian Association for Regional Cooperation (SAARC) and hence is a representative sample of multiple manifestations of trends in the South Asian region. The presence of both China and Vietnam in this category allows a comparison between nations with similar land use and economic levels, but whose sizes and governance systems are vastly different. The use of different categories allows comparisons between countries whose economic conditions are different.

Time series plots for the various categories of countries were prepared for population, GDP, and GDP per capita in order to examine growth rates in the consumption base. Although these plots end in 1987, this does not limit the quality of inference since this data is used to construct patterns that extend beyond that date. Indeed, more recent data corroborates our projections (World Resources Institute website, 2000). Agricultural land use was examined through a per capita food production index, including fertilizer and pesticide application, cultivable land area, and forests. Additionally, related patterns pertaining to soil quality and fertility were constructed from a soil science knowledge base. The following observations were made with respect to the growth of the consumption base and the conditions of agricultural land.

Growth of the Consumption Base

Figure 8.1 (a, b and c) shows population and GDP growth in the three categories of countries selected for the analysis. Considerable population growth is shown over the three decades covered by the data in all categories, although this growth is much higher in the low-income countries of South Asia. GDP growth is the highest in the middle-income countries, while they are comparable between the high- and low-income countries. Consequently, as shown in Figure 8.2 (a, b and c), GDP per capita has grown at comparable rates in the high- and medium-income countries due to moderate population growth in the former and high economic growth in the latter. In comparison, high population growth rates and moderate economic growth have led to stagnation in GDP per capita in the low-income countries. (All monetary data has been converted to US dollar equivalents. Food production indices are based on 1979-81=100. Some of the time series required unit conversions and reconciliation of contradicting numbers. In all cases, more recent numbers were preferred over older data. Some of the missing data cells were computed through interpolation between existing entries).

According to the projections of the United Nations Centre for Human Settlements (UNCHS) shown in Figure 8.3 (a, b and c), tremendous growth has also occurred in urban populations across the board and the high growth rate is expected to continue, although these rates are projected to taper off in the high-income countries. On the other hand, rural populations have shown stagnating or declining trends in higher income countries and may be expected to decline further in the future.

Figure 8.1
Population and GDP growth

a) High Income Countries

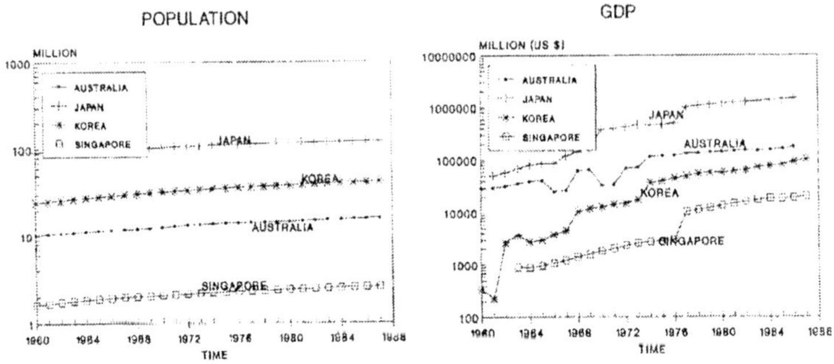

b) Middle Income Countries

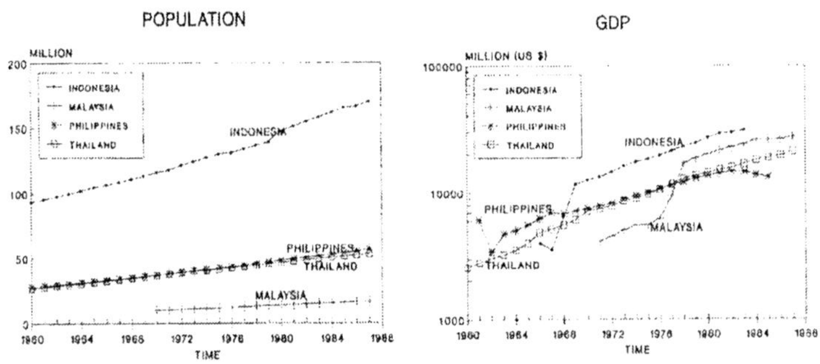

c) Low Income Countries

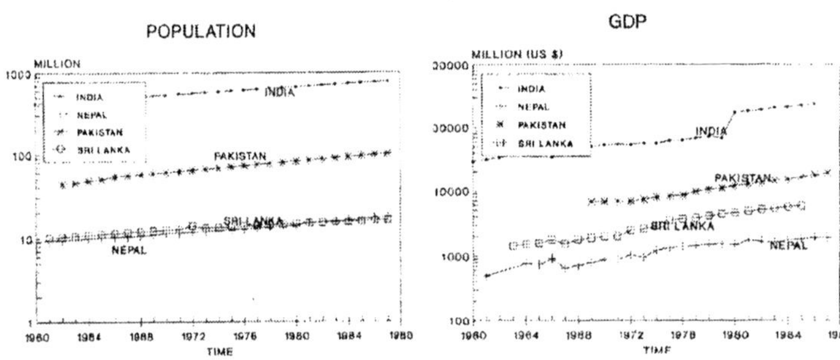

Figure 8.2
GDP per capita growth

a) High Income Countries

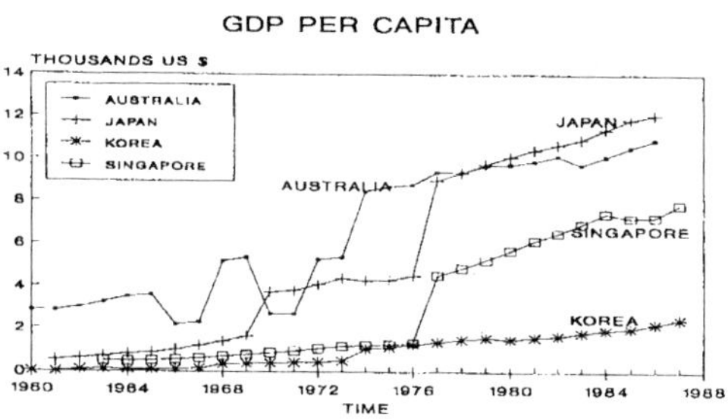

b) Middle Income Countries

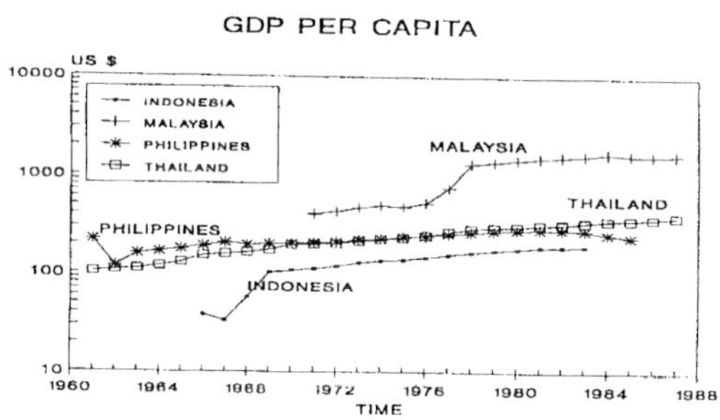

c) Low Income Countries

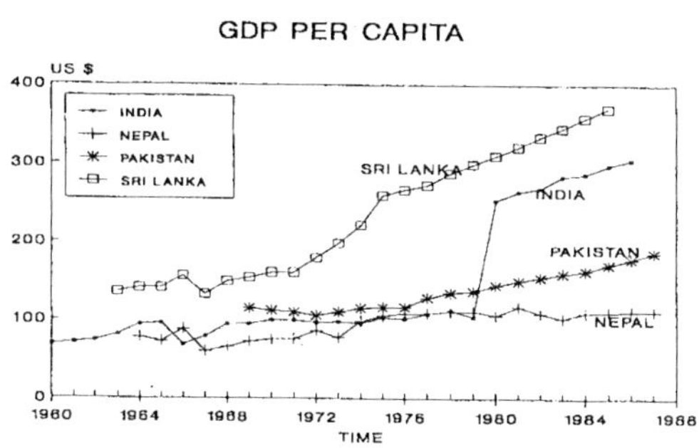

Figure 8.3
Projections for total and urban population

a) High Income Countries

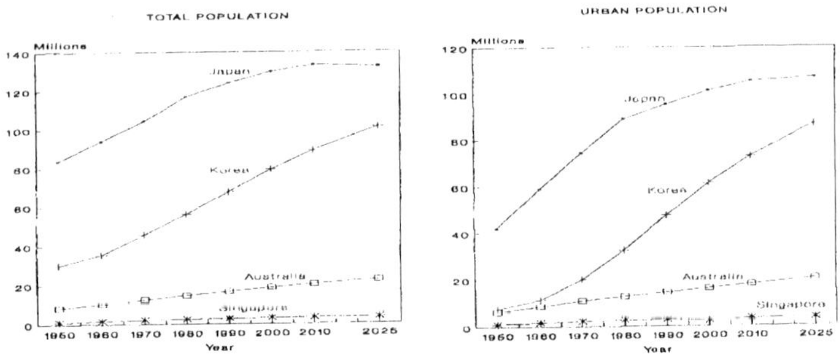

b) Middle Income Countries

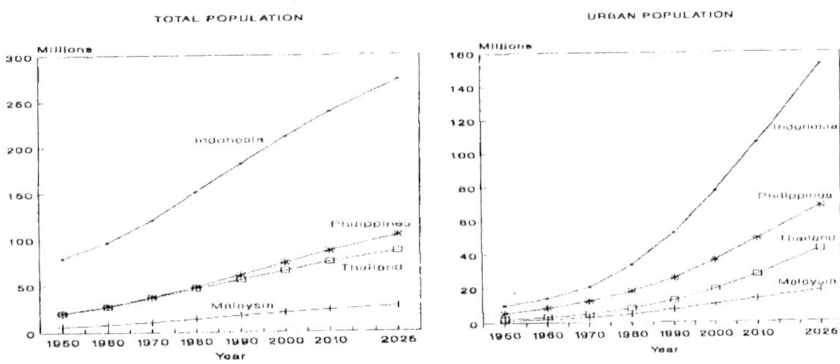

c) Low Income Countries

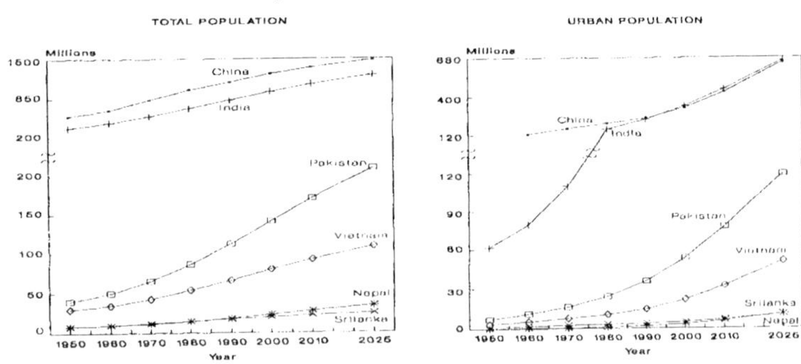

However, due to the overall momentum of population growth, rural population levels have risen significantly in the medium- and low-income countries, but are expected to taper off and begin to decline over the second decade of the twenty-first century. Figure 8.3 also shows that the total population is expected to continue to rise in all countries well into the twenty-first century. However, the rates of projected population growth are negatively correlated with the levels of income, where lower-income countries experience consistently higher rates of total population growth and urbanization (UNCHS, 1987). In all cases, there is growth in the consumption base originating from two sources: growth in population and expansion in economic activity.

A significant side effect of the expansion of the economic base is the rise in urbanization, which is necessitated by technological choices outlining the level of production activity necessary to achieve economies of scale.

However, as previous chapters in this volume have pointed out, the concentration of human activity tends to also create concentrated doses of pollution, which the environment is unable to assimilate. The results of this are easily observed in the falling water table, the pollution of groundwater aquifers, and acid rain, all of which adversely effect agriculture. In addition, urbanization often encroaches on prime agricultural land, which reduces production potentials, especially in Asia where only marginal lands remain uncultivated. All of these pathologies are particularly prevalent in the South Asian context.

Condition of Renewable Agricultural Resources

In this context, renewable resources include agricultural land and forests, which have traditionally met the food, fuel, and timber needs of South Asian society. Figure 8.4 (a, b and c) shows past trends in food production per capita and agricultural land per capita in the members of each of the three sample categories. A logarithmic scale is used to represent land per capita in high-income countries in order to accommodate the variety of data. The food production index is not comparable between nations due to differences in the criteria used to calculate the base figures. The index represents only an internal measure of the changes in food availability in each country.

Figure 8.4
Food production per capita and agricultural land per capita

a) High Income Countries

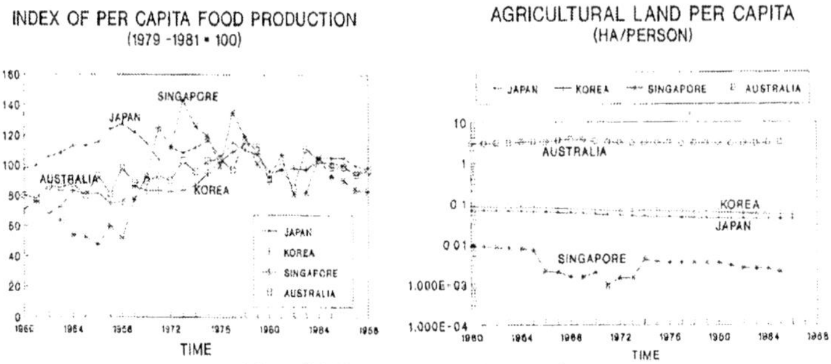

b) Middle Income Countries

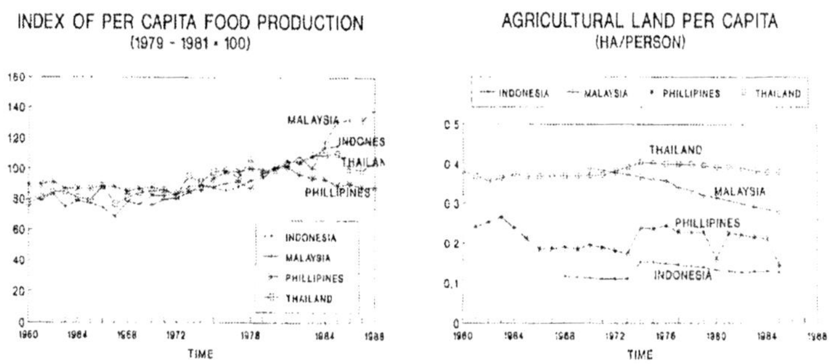

c) Low Income Countries

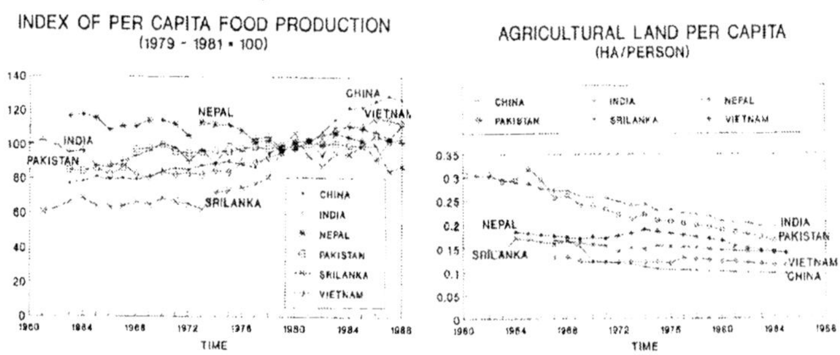

Figure 8.5
Fertilizer and pesticide application

a) High Income Countries

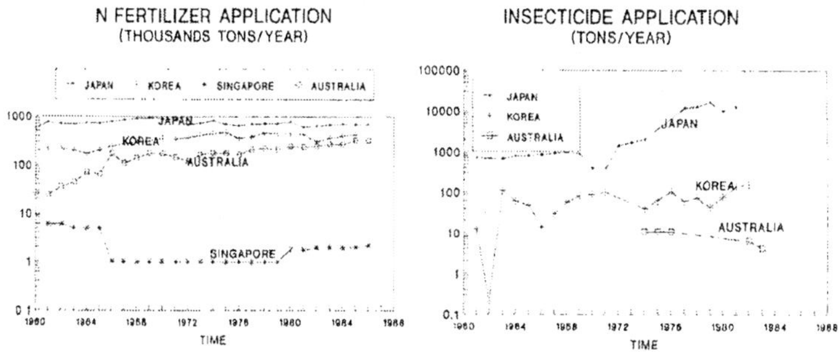

b) Middle Income Countries

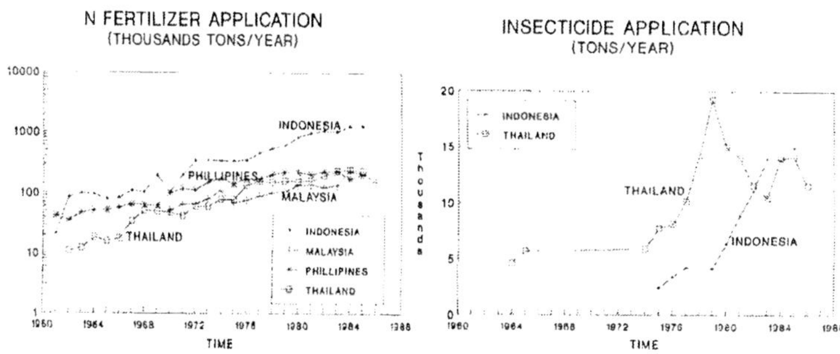

c) Low Income Countries

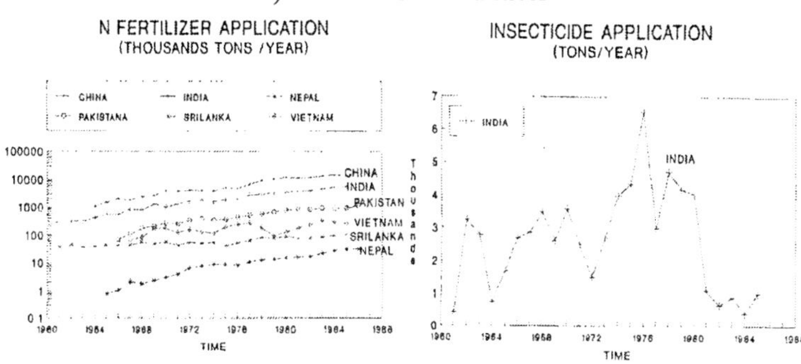

Figure 8.6
The competition between cultivable and forest land

a) High Income Countries

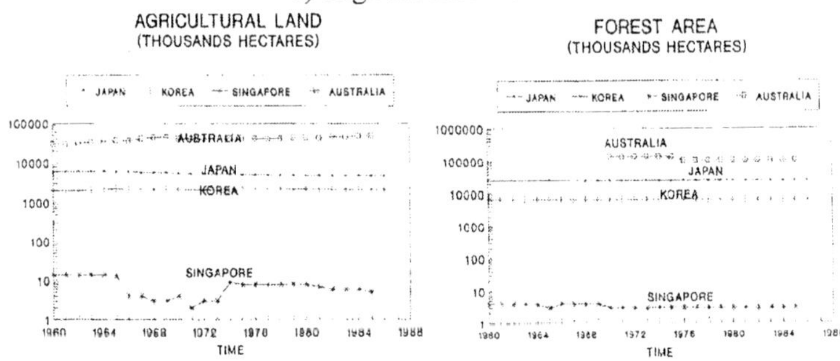

b) Middle Income Countries

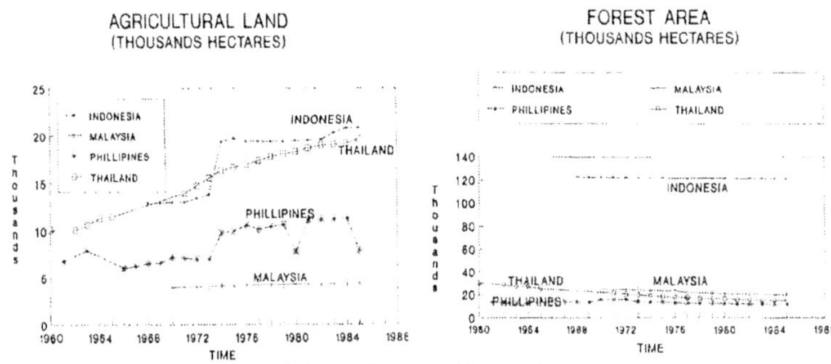

c) Low Income Countries

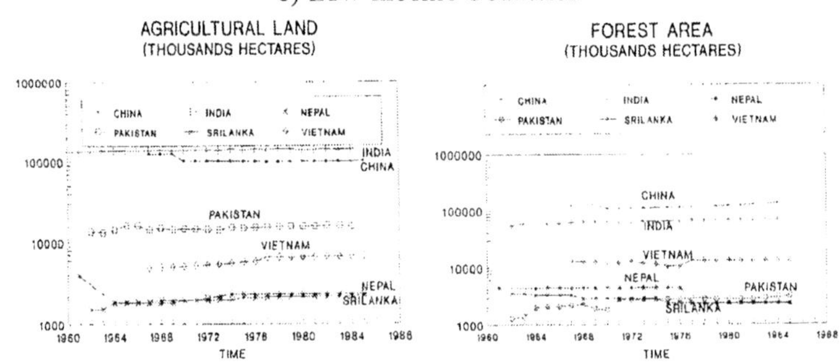

Some autonomous jumps also appear in the data, since it has been constructed from many sources, which although it has mostly been published by the UN, contains inconsistencies in the definitions used to represent the various categories of data. For articulating the food security problem, however, long-term patterns rather than the numerical values of the time series are compared. Hence, the data's shortcomings can be tolerated.

It is observed that food production per capita is rising in all cases in spite of considerable population growth, while agricultural land per capita is declining, except in Australia, where it has been possible to maintain it at a steady level. This indicates that increases in food production have been obtained largely through increasing the intensity of cultivation and application of chemical fertilizers and pesticides. Indeed, as indicated in Figure 8.5 (a, b, and c), fertilizer application (in logarithmic scale) has drastically increased over the past three decades in all countries studied. The application of pesticides (in logarithmic scale for high-income countries) also seems to have increased in the countries for which data is available. The data on pesticides, however, is inconsistent since in some cases it refers only to DDT while in others it covers all pesticides.

Despite the increases in yield, the quantity of arable land has not increased much in most of the countries studied, except in Australia, where it has been possible to commission large tracts of unused land. This is shown in Figure 8.6 (a, b, and c) (where logarithmic scale is used for high- and low-income categories). It is observed that, in general, where the amount of cultivable land did increase, it was at the expense of the forested area, which is already very small in the countries that have a static amount of land used for agriculture.

Some jumps again appear in the plotted data due to changes in the definitions of forest area and agricultural land categories, which do not stand up against the underlying pattern of change. However, deforestation not only reduces valuable resources in terms of timber and wood for fuel, it also causes soil erosion, water loss, flooding, drought, desertification, and silting of irrigation reservoirs, depending on the specific function the forest performs in the ecological system (Bowonder, 1986). In spite of this knowledge, about half of the forest area in the developing countries was cleared between 1900 and 1965.

At current rates of deforestation, the rest is likely to disappear in 50 years (UNESCAP, 1986).

Figure 8.7
Declining food per capita trends in Nepal and Bangladesh

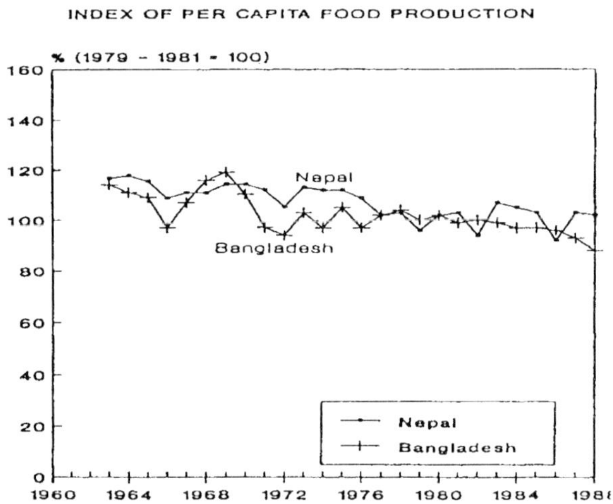

Excessive use of land resources has also been known to depreciate soil quality. Soil degradation has occurred, in the countries studied and elsewhere, because of erosion, chemical deterioration, loss of texture, waterlogging, and salinity, all resulting from efforts to intensify agricultural activity (Bowonder, 1981). Given the overuse of land resources, the per capita food production index may be expected to decline across the board in the future. A similar variety of aggregate trends for agricultural production at the continental level has been reported by the World Resources Institute (WRI website, 2000; http://www.wri.org).

Juxtaposing these results with accounts of resource degradation in individual South Asian countries, as provided in earlier chapters in this volume, produces rather disturbing trends. Clear trends toward decline have already appeared in Nepal and Bangladesh, as shown in Figure 8.7, and given the similarities in their resource use patterns and policies, these trends are likely to be repeated elsewhere in South Asia.

Agricultural Land Use Patterns

The observed trends in the data taken from a geographically, economically, and politically diverse set of nations show that in all cases, increases in agricultural production have been achieved by making an intensive use of land resources treated as investments rather than environmental systems. Moreover, these are clearly private gains, whether pursued by individuals or collectives. It is also quite evident that the expansion of land for agriculture has been achieved through the consumption of forests. This is another environmental system that is important for the maintenance of agricultural land as a sustainable resource, but that is also viewed by those involved in agriculture as an 'unused' endowment. A variable implicit in the above description is soil nitrogen and humus, which are dependent on the intensity of cultivation, crop rotation, and soil management. Soil fertility has evidently been upheld by the application of fertilizer, but it will eventually decay if the pressure on soil continues.

Excessive use of agricultural land has been known to depreciate soil quality. Soil degradation has occurred in many countries due to erosion, loss of soil nutrients and texture, waterlogging, and increased salinity, all of which are usually caused by intensive land use (Bowonder, 1981; also see Asthana and Shukla, and Rahman, et al. in this volume). Both centrally-planned and market-economy nations have encountered these problems. Market economies tend to externalize private costs to the environmental commons, since decisions made by individuals often consider the environment a free resource. The centrally-planned economies, likewise, fail to internalize environmental costs because incentives are provided to managers to boost production, not preserve the environment. Furthermore, since the resources placed in the hands of the managers do not take the scarcity value, opportunity cost, or real price into account, the cost of using resources is essentially irrelevant (Chandler, 1987).

System Relationships in a Non-sustainable Food Production System

The food production system of these countries can be characterized by the feedback loops shown in Figure 8.8. Food fulfills the nutritional needs of the population, hence food sufficiency is definitely related to the average life expectancy and also to population growth. Food production depends on cultivated land and harvest biomass, which is increased by intensive cultivation, among other factors. As a result, non-harvest biomass that maintains soil humus is decreased. An increase in population expands the food consumption base. Consequently, food production is stepped up through intensive land use, high yielding seed varieties, and extensive irrigation, all of which degrade land in the long run from decay in humus and nitrogenous matter. Yield may also be increased or sustained through investments in land improvement, which is often a last resort after much damage has already been caused. For simplification, forests and wastelands are placed in an aggregated category. Pressure for food production may lead to the clearing of forests and the commissioning of already-cleared lands and wastelands to be used for agriculture. Given the current levels of forestation and protectionist policies, if implemented, this policy would mostly be directed toward wastelands.

These relationships were incorporated into a formal system dynamics model that was experimented with in order to understand the implications for the future (Sterman, 2000; Richardson, 1997). The model designates three subsystems: population, food production, and ecology. The structure of each is discussed in Bach and Saeed (1992). The model was parameterized for the ten countries in the medium- and low-income categories, with the low-income countries also including the SAARC region. The nations in the high-income category seemed to exercise technological options that were not purely pressure-based. It was therefore not considered appropriate to apply the model to those countries. Figure 8.9 summarizes the results of the simulations of the model with ten parameter sets representing the ambient conditions in the countries that lie in the middle- and low-income categories.

Figure 8.8
Key feedback processes in the food production system

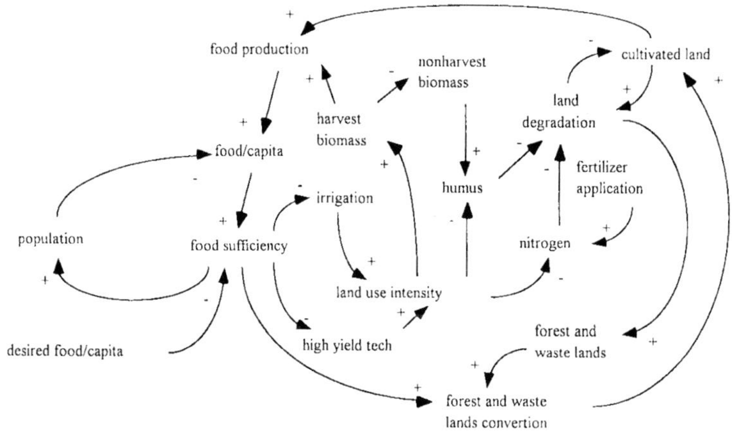

It is observed that although food production per capita has been sustained in most cases by an increasing input of capital and cultivation intensity, this situation is likely to reverse itself fairly soon. Asthana and Shukla seem to validate this very finding in the case of India elsewhere in this volume. Moreover, land quality may be expected to deteriorate without stopping, which could cause sudden declines over the next two decades unless concerted land conservation and reclamation efforts (not included in the model) are put into place. Countries affected most are those with relatively high population growth rates, intensive land use levels, and low forest coverage. In essence, most of South Asia is particularly at risk. Similar trends obtained with ten parameter-sets representing the low and medium countries of the sample also show that the system is parameter insensitive. Hence, these inferences can be extended beyond the analyzed set.

Since the magnitude of deterioration in the land quality index shown in the simulations does not take existing land management programs into account, it indicates only the volume of effort needed to overcome the deteriorating trends. Yet the results of the simulations are borne out by the evidence from earlier chapters (especially those on Pakistan and India) and the experience of many developing countries with large agricultural economies where fertile agricultural lands have already been extensively damaged due to waterlogging, increased salinity, and erosion of topsoil. Some of these countries have already instituted extensive land management programs (Allauddin and Tisdell, 1988).

Figure 8.9

Model behavior for parameter sets representing countries in low and medium income categories

Low Income Countries

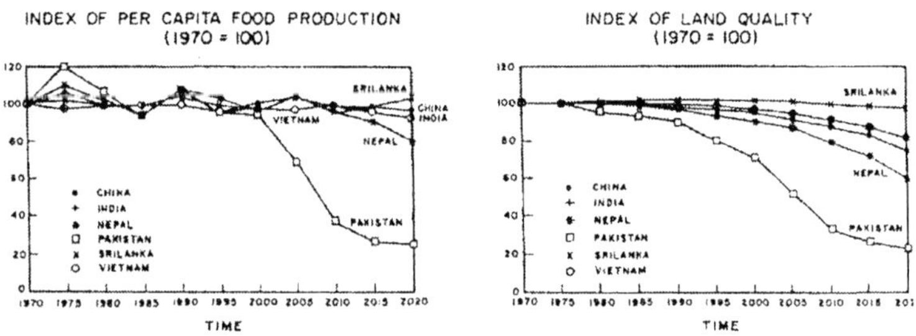

Middle Income Countries

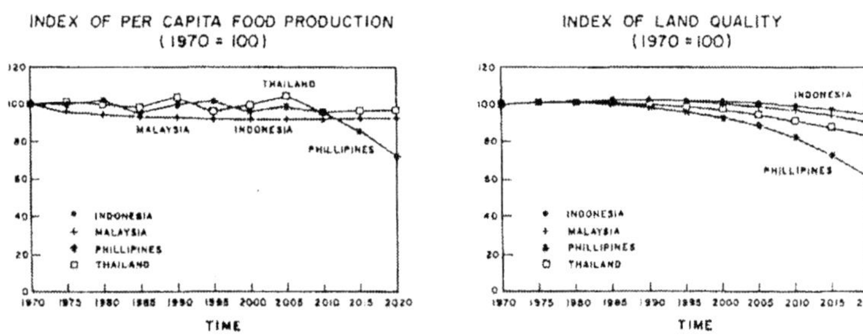

The trends observed in historical data provide persuasive evidence that, over the course of agricultural development, increases in output have been obtained by consumption of resources, which has in turn reduced food security. The projections made with the help of a model incorporating organic cause and effect relationships show that this process has gone far enough and cannot be sustained for long into the future.

Generating Sustainable Agricultural Policies and Practices

A decision-making process cannot be divorced from its organizational context and it would be impossible to go beyond speculation on issues concerning a commons unless an institutional framework accepting responsibility for the commons is created. Sustainable agricultural technologies in the form of energy-efficient crops and soil-preserving cultivation methods are now widely available. However, neither an adequate awareness of the environmental issues nor an institutional framework to take responsibility for the environment is in existence in the polities of South Asia. Hence, sustainable technologies have a poor chance of being adopted. This section seeks to highlight key social and institutional policy directions that might be pursued to help inspire responsible attitudes toward environmental systems, ideally leading to the adoption of sustainable technologies at the governmental level.

Changes in Social Policy

Despite some efforts at reform, agricultural land ownership is concentrated in only a few hands in most countries of South Asia. A large proportion of the small farmers remain landless or with only subsistence holdings. Hence, farmers do not have a stake in the preservation of the land they till on a rental or sharecropping basis. Even when they do own the land they till, it is often in small tracts that must support large self-employed families with a much higher priority on immediate consumption than on ecological preservation. Finally, there might exist an awareness of the environmental issues and a commitment to preservation, but there are no institutional mechanisms for mobilizing individual initiatives into the maintenance and preservation activities that call for a collective effort. Thus, the following social reforms must precede any sustainable agricultural development initiative:

- Fiscal instruments and direct intervention should be introduced for transferring absentee-held (both state and private) agricultural resources directly to the cultivators so they have a personal stake in the preservation of these resources.

- Extension services should be provided to help cultivators increase their income so conservation can take priority over consumption.
- Community organizations serving the collective interests of small farmers should provide incentives to address the issue of the commons.
- Infrastructure development should be pursued for increasing social mobility in the rural areas, thereby making it possible for the surplus labor to move into other production sectors, reducing pressure on land resources.
- Fiscal policies should be designed and implemented for creating a crop mix that emphasizes food self-sufficiency on a long-term basis rather than income generation from export of primary products.
- Population control should be increased as a measure to contain the demand for food.

Institutional Policy Directions

Current approaches to national planning and policy design in South Asian countries have often entailed fragmented efforts undertaken in the compartmentalized wings of government, which are ultimately assembled into a statement of targets by the central planning organizations. This process can rarely identify institutional mechanisms for creating change. To create environmentally-sound development policies, it would be necessary to reorient planning activities so that it is possible to design institutional instruments of change alongside the preparation of target statements. This calls for rethinking the structure of the planning institutions as well as education of socio-economic planning along the following lines:

- The information base concerned with public policy should be expanded to include geological and environmental information as well as models of organic relationships. These may be experimented with to generate environmentally rational scenarios. This would require the reorganization of planning, environmental, scientific, and technological functions of the government, so that their actions are integrative rather than compartmental as is currently the case in most South Asian countries.

- Since environmental issues may clash with powerful corporate and political interests, environmentally-conscious institutions should be created in the form of private, non-government, and government interest groups. A number of non-governmental interest groups and organizations that pursue environmental agendas have recently emerged in South Asia. These groups should also be given legislative protection and a role in decision-making on public agendas.
- Agricultural research should redefine its objectives and seek development of seed varieties and cropping patterns for low-energy input and robustness. Agricultural research in South Asia has accomplished the opposite by creating the so-called 'green revolution technologies' that require high-energy input to deliver high-yield, soil-draining, and ultimately vulnerable crops.

Conclusion

This paper has attempted to illustrate ways in which the pervasive indifference toward the conservation of the environment on the part of the public and governmental institutions is creating dangerous long-term trends depleting agricultural resources in Asia at large, and in South Asia in particular, and that these trends pose a serious threat to food security. Such attitudes arise from vantage points, both in private and public decisions, that emphasize increasing production in the short term by externalizing costs on the common environment, whose maintenance is not a direct responsibility of any of its beneficiaries. The increased availability of production, however, also causes growth in the consumption base, thereby limiting availability and requiring further increases in production. The environmental abuse of this downward spiral has, by now, created serious deteriorating trends and seriously threatened sustained agriculture. If these deteriorating trends are to be reversed, the policies for the private and public decisions concerning agriculture must clearly be expanded to incorporate information about stocks and organic relationships in the interaction between human decisions and the environment. This is not a simple task and will require major institutional reform, in both the planning and use of information and in changing existing attitudes.

The themes presented in this book clearly demonstrate that human vulnerability has increased particularly in South Asian countries due to the implementation of a developmental agenda that addressed symptoms rather than underlying causes of problems. Table 8.1 shows the selected developmental problems and the broad policies implemented to alleviate them. The first issues perceived to threaten human security were hunger, poverty, and insurgence. Since these problems were taken as given, the natural response to alleviate them was to facilitate intensive agriculture so more food could be produced, to foster economic growth so personal income could be increased, and to strengthen the internal security and defense infrastructures so insurgent groups could be suppressed.

The common denominator of those policies was that they attributed the existing conditions to outside factors, as if they came about as autonomous acts of fate. The assumption was also made that the system is static and non-self-regulating. Thus, it was expected that directly attacking the symptoms of the problems would help alleviate them. The problems that followed were numerous and in most cases included a continuation of the existing problems that the short-term solutions failed to solve.

Thus, food shortages have continued but are now also accompanied by land degradation, depletion of water aquifers, the threat of large-scale crop failure due to a reduction in crop diversity, and a tremendous growth in population. Poverty and disparities in income between the rich and the poor have in fact shown a steady rise, accompanied by unprecedented debt burdens, extensive depletion of natural resources, and degradation of the environment. Political insurgence and threats to peace have intensified together with burgeoning expenditures on internal security and defense. This has stifled the development of social services and human resources and has also created opportunities for authoritarian governments with little commitment to public welfare.

Table 8.1
Developmental problems, policies implemented to address them, and subsequent problems

Initially perceived problems	Policies implemented	Subsequently experienced problems
Food security	▪ Intensive agriculture *Land development* *Irrigation* *Fertilizer application* *Use of new seeds*	▪ Land degradation ▪ Depletion of water aquifers ▪ Vulnerability to crop failure ▪ Population growth ▪ Continuing/increased vulnerability to food shortage
Poverty	▪ Economic growth *Capital formation* *Sectoral development* *Technology transfer* *External trade*	▪ Low productivity ▪ Indebtedness ▪ Natural resources depletion ▪ Environmental degradation ▪ Continuing/increased poverty
Insurgence and threats to peace	▪ Spending on internal security and defense infrastructure ▪ Limiting civil rights	▪ Poor social services ▪ Poor economic infrastructure ▪ Authoritarian governance ▪ Continuing insurgence ▪ Increased threats to peace

The subsequent problems experienced are more complex than the initial problems were, and have recently drawn expressions of concern at the global level. However, it is questionable whether an outside hand at the global level would alleviate them. This is evident from the failure to formulate and enforce global public policy in spite of active participation by national governments, global agencies like the UN, the World Bank, the World Trade Organization, and advocacy networks sometimes referred to as the 'Civil Society'. This failure can largely be attributed to the lack of a clear understanding about the roles of the actors who precipitated those problems and those whose motivations must be influenced to turn the tide.

Developmental interventions must adopt a problem-solving approach in a scientific sense if they are to enhance human security. Under this approach, a problem should be defined as an internal behavioral tendency found in a system and not as a snapshot of existing conditions. It may represent a set of patterns, a series of trends, or a set of existing conditions that appear resilient to policy intervention. In other words, an end condition alone must not be considered the problem. The complex pattern of change preceding the end condition would, on the other hand, represent the problem. The solution to a recognized problem should be a scientific one, dedicated to the understanding of the underlying causes. A development policy should then be adopted as a change in the decisions made at the governmental level designed to reform a problematic pattern to one that is acceptable.

Chapter Nine

Water & Security: A Reexamination

RAMASWAMY R. IYER

THIS PAPER IS concerned with ideas, approaches and attitudes; not with numbers. Taking it for granted that there will be growing pressure on the available supplies of fresh water in this subcontinent, this paper will argue for changes in the prevalent ways of thinking with a view to minimizing that pressure and the concomitant conflicts.

A Cautionary Preamble

At the outset it seems necessary to sound a note of caution regarding the tendency, now fairly common, to talk about conflicts over scarce natural resources or environmental concerns in the language of 'security.' Conflicts can certainly arise, and have arisen, over these matters, and they can become acute, perhaps even violent; however, this writer does not subscribe to the currently fashionable theory of 'water wars', and considers the analogy with oil most misleading. What is most worrisome is the importation of the notion of 'security' to this arena (Iyer, 1998 *for the author's fuller explanation of this view*).

Before we explore that issue, however, let us first get one obvious point out of the way. It is of course possible to use a natural resource or an environmental factor as weapon of war or an instrument of political pressure. To give some fanciful hypotheses, a country may try to poison the water source of another, or bomb a dam in an enemy nation to create devastation, and so on. These are straightforward security concerns in the conventional sense; the fact that natural resources or environmental factors are used as the means of inflicting damage instead of the usual weapons is merely incidental. Again, where a river runs along or marks the boundary between two countries, there may be security concerns relating to that border similar to those relating to a land border. These are not the kinds of things that we are concerned with here; what we have in mind is the possibility of conflicts arising from scarcities of natural resources, their pollution or contamination through use, environmental degradation, and so on. It is in relation to such conflicts, existing or potential, that the language of security seems to be both dubious and dangerous.

Those are strong words that need to be justified. First, why 'dubious'? Because there exists a fundamental confusion between different senses of the word 'security'. On the one hand, there are the familiar security concerns of defense ministries, military leaders, external and internal intelligence agencies and the internal police and paramilitary forces. Their worries include the capabilities of the armed forces of other countries, including their weaponry and technology; the dangers of espionage, subversion, terrorism and destabilization; internal armed insurgency, riots and civil wars; and so on. That is one sense of the word 'security', perhaps the dominant sense. On the other hand, we also use it in the context of arrangements to look after the safety of industrial establishments, offices, shops, homes or individuals. In yet another context we talk about 'social security'. In all these uses of the word, there are two common components in the meaning, namely *risk* and *protection*, however they are of very different kinds in very different categories.

When we talk about 'food security' or 'economic security', we are closer to the usage in 'social security' than to 'security' in the military or police sense. Under certain circumstances the non-military and military senses of 'security' may tend to merge. However, it is useful to maintain the distinction between the two senses. The mere fact that the same word – 'security' – is used in diverse contexts should not mislead us into thinking that we are talking about the same thing, thereby transferring concepts and approaches from one context into another.

Secondly, why 'dangerous'? Because 'security' in the military sense is essentially a negative and adversarial concept. Its perception and analysis are composed of threats and the formulation of counter-measures, defensive or offensive. 'Security', in this sense, is always *against* an adversary. Quite frequently, it is a nationalistic notion (though there are instances of regional security arrangements); boundaries and barriers are important. All this is far removed from the language of harmony or cooperation.

When we are considering scarce natural resources or environmental concerns, we are in an altogether different universe of discourse. Here we have to think in terms of cooperation, harmony and collaboration: in terms of natural systems, not national boundaries. Conflicts exist, of course, or may emerge, and they have to be resolved; but we would do well to use precisely that language – namely, 'conflicts' and their 'resolution'. When we begin to talk about these matters in the language of security, the entire tone of the discussion changes into an embattled, defensive, distrustful, adversarial mode. A further danger is that the wrong actors enter the arena. The use of the terminology of security makes the 'security community' feel that this is *their* area, and this makes the processes of conflict-resolution more, and not less, difficult.

We must, of course, recognize the commonality of certain natural resources or ecological systems to a number of countries, perhaps even a region, as well as the fact that what one country does or fails to do in relation to those resources or that system within its own borders can have consequences beyond them. From that recognition we can proceed in one of two different and divergent ways. We can proceed towards treaties, agreements, bilateral or multilateral or regional arrangements or institutions – in short, towards harmony and cooperation. Or we can talk in terms of security, threat perceptions, defensive measures, and so on, and promote disharmony. In brief, incorrect terminology provides an incorrect framework, brings the wrong people in and pushes us in the wrong direction.

Some may say, "we are not using the word 'security' in a narrow, adversarial sense. We are, in fact, trying to enlarge the concept to cover a wider range of concerns, and to give it a *positive* slant." But why torture the concept and force it to take on meanings that are not natural to it? However hard we try to sanitize the word and replace its negative, adversarial connotation by a positive, consensual one, a residue of its original sense and undertone will remain, which will be carried into the newer applications. Our efforts to widen the mental horizons of the traditional security community and to sensitize them to newer concerns may or may not influence *their* thinking, but a borrowing of their terminology will surely influence *ours*.

The Area of Concern

Having entered those reservations, it becomes obvious that one cannot hope to change the current usage. Let us then proceed to deal with the issue as posed. In the context of water, we are concerned mainly with the South Asian sub-continent. Island countries such as Mauritius and Sri Lanka are not linked to the water systems of the mainland, therefore, we can exclude them. For the present, at any rate, we can leave out China too, for the reason that it is physically separated from the South Asian landmass by mountains and plateaux. The Brahmaputra (or Tsangpo) flows largely through uninhabited or sparsely inhabited areas until it crosses into India. There were reports some years ago that China was planning to divert the Brahmaputra northwards with the aid of nuclear explosions. It is not clear whether there are in fact any such plans. If there are, that would be a matter for serious concern on the part of both India and Bangladesh. For the time being, however, there are no water-related issues between China and India. We are, therefore, talking about Pakistan, India, Nepal, Bhutan and Bangladesh, or to put it in terms of river systems, the Indus system in the west and the Ganga-Brahmaputra-Barak/Meghna (or GBM) system, or systems, in the east (there is a difference of views between India and Bangladesh on whether these three rivers constitute one system or three).

Both in the west and the east, the river systems were cut across by the lines of the partition of 1947, transforming them into trans-boundary systems requiring inter-country understandings. It will be noticed that the reference is essentially to the northern half of the subcontinent. The southern peninsular part is unrelated to the Himalayan river systems, and while there have been and continue to be conflicts over the southern rivers (Godavari, Krishna, Cauvery and others), there are no international dimensions to such conflicts. International dimensions could arise if any attempts were made to divert water from the Himalayan rivers to the peninsular ones. There has been some vague, unrealistic talk of such possibilities but it seems very unlikely to become a serious proposition in the foreseeable future.

Against that background, what can 'security' mean in relation to water in this area? We can best answer that question not by trying to define the term, but by trying to capture the kinds of concerns that lie behind its use. Approaching the subject in this manner, we can identify three broad categories of 'security' issues:

- Water needs for diverse purposes: availability, adequacy, reliability, dependence, vulnerability, etc.
- Floods: need for mitigation, management, and damage-minimization.
- Water quality problems: prevention or control of pollution and contamination.

If we identify the problems in this manner, what needs to be done is clear enough:

- Understandings on water-sharing in common river systems. (In referring to rivers that flow through more than one country, one has to use words such as 'common' or 'shared,' but even these can give rise to controversy. For example, at one stage there was a strong reaction in some quarters in Nepal to the use of the term 'common rivers' in a Memorandum of Understanding between the Prime Ministers of Nepal and India!)

- Cooperation in establishing and operating effective and timely information and warning systems in regard to flood flows, and in disaster-preparedness and damage-mitigation; a coordination of coping strategies; a sharing of experiences in this regard.
- The institution of common standards on water quality, and understandings in regard to the maintenance of water quality across borders.

It will be noticed that 'energy security' is not referred to above. Apart from the fact that there is another paper dealing with that subject, it seems to this writer that with hydro-electric power – which is what we are concerned with in the context of water – the question is not so much one of security or insecurity as one of a potential to be utilized.

Floods

Floods and the proper response to them is a vast, complex and controversial subject that will need a separate paper. For our present purposes what we need to note is that floods have been a part of the experience of the countries in this sub-continent from time immemorial and will continue to remain so in the future. The extent and duration of flooding and the damage it causes are influenced by both natural phenomena and human activities. The natural causes include prolonged and heavy rainfall, intense precipitation during a limited period, sudden cloudbursts, the breaking of temporary barriers created by landslides or earthquakes, the occasional simultaneous occurrence of high floods in different river systems (e.g., the Ganga and the Brahmaputra, normally have their peaks at different times), the further synchronization of this with tidal bore conditions in the estuary, as happens occasionally in Bangladesh, and so on. Human-induced causes include the deforestation of hillsides, structures impeding river flows, human settlement and economic activity in the flood plains of rivers, drainage congestion, and so on.

Some of the efforts at flood control (for example, embankments and dams) have sometimes themselves caused further problems. There is a growing realization that 'flood control' is not a feasible proposition; that embankments have often proved to be a remedy worse than the disease; that large dams are not often planned with flood moderation as a primary aim, and even where they are, the competing claims of irrigation and power-generation often override the flood-moderation function; that dams, while they might moderate flood flows to some extent under normal conditions, might aggravate the problem if water has to be suddenly released in the interest of the safety of structures; and that what we must learn to do is not so much to control floods as to cope with them when they occur and minimize their damage.

The most important element here is timely knowledge and preparedness: this is the meaning of 'security' in this context. Governments and individuals need to know how soon a flood is likely to arrive and what its magnitude is likely to be. They can then take appropriate measures for the prevention or minimization of hardship, loss and damage, and for relief where necessary. A vast, well-equipped, technologically advanced network of stations is needed which would observe and analyze precipitation and water flows, draw conclusions about them, and provide instant (or 'real time') communication of such information and predictions to downstream areas and countries. This is an important area for inter-country cooperation. Not only knowledge but experience and lessons need to be shared, and coping strategies exchanged and coordinated. Modest beginnings have been made in these directions but much remains to be done. (Adhikary, et al., 2000; *Seminar*, 1999).

Water Quality

Water quality is another important area of concern. As previous chapters have highlighted, there has been rapid and grave deterioration of water quality in both surface waters and groundwater aquifers in all the countries of South Asia. All of them have pollution control laws and institutions, but these have not been able to prevent the growing pollution and contamination levels of water sources and systems, which in effect makes much of the available water resources unusable. This is, in fact, as great a threat to security as the 'scarcity' about which alarm bells have been ringing. Much of the action here – prescription and continuous review of standards, their enforcement, remembering the cumulative impact of individual clearances and permits, making the polluter pay, adopting and moving towards clear, time-bound goals in regard to desired water quality – lies within each country. However, pollution and contamination do not respect political boundaries. The countries of the region have to agree on common standards and on transboundary water quality protocols. Conflicts have arisen in the past over water-sharing, but water quality may well become the focus of even sharper conflicts in the future unless clear understandings are reached between nations and appropriate institutional mechanisms are provided for ensuring compliance with them. (Adhikary, et al., 2000).

Having very briefly dealt with those two aspects of security, let us now turn to what is generally perceived as the principal issue: water sharing.

Water Sharing

India-Pakistan

There is already an understanding between India and Pakistan insofar as the Indus system is concerned: the Indus Water Treaty of 1960. The water sharing under the treaty was quite simple: the three western rivers (the Jhelum, the Chenab and the Indus itself) were allocated to Pakistan, and the three eastern rivers (the Ravi, the Beas and the Sutlej) were allocated to India. Certain restrictions were placed on India as the upper riparian. it was not allowed to build storages, except to a very limited extent, on the rivers allocated to Pakistan. Restrictions were also imposed on the extension of irrigation development in India. In Pakistan, the lower riparian, there were some relatively less-significant restrictions. There were also provisions regarding the exchange of data on project operation, extent of irrigated agriculture, and so on. The treaty further mandated certain institutional arrangements: there was to be a permanent Indus Commission consisting of one Commissioner each for India and for Pakistan, and there were to be periodical meetings and exchanges of visits. Provisions were included for conflict resolution: differences, if any arose, were to be resolved within the Commission; if agreement could not be reached at the Commission level, the dispute was to be referred to the two governments; if they also failed to reach an agreement, the treaty provided an arbitration mechanism. The settlement also included the provision of international financial assistance to Pakistan for the development of irrigation works for utilizing the waters allocated to it, and India paid a sum of approximately 10 billion rupees to an international fund for this purpose.

The Indus Water Treaty has been working reasonably well despite the difficult political relationship between the two countries. In fact, it was not abrogated even during periods of war. The Indus Commissioners meet regularly, and the working relationship between the engineers at the Commission level is very cordial. Differences do arise from time to time, but these usually get resolved within the framework of the treaty. At present there is an unresolved dispute regarding what is known as the Tulbul Navigation Project (or the Wular Barrage Project). The inter-governmental talks on the subject have not so far been successful, but there is no reason to believe that an agreement will not eventually be reached.

It could be argued that the division of the river system into two segments, one for Pakistan and one for India, was not the best method of sharing, and that there should have been a sharing on all the rivers, or a joint integrated planning and management of the entire system. However, the ideal may not always be practicable, and the Treaty represents an agreement arrived at by the two countries after prolonged negotiations. What is important, and a matter for gratification, is that the treaty has actually been working in a reasonably satisfactory manner. There are of course problems in the Indus basin, such as waterlogging, salinization, and drainage congestion, but these cannot be attributed to the treaty; they are the result of bad water management and irrigation practices in both countries.

It follows from the above that water security will not be an issue between India and Pakistan, so long as the Indus Treaty continues to work. Population growth, the pace of urbanization, and the progress of what we call 'development' may indeed impose increasing pressures on the finite availability of water, but this will call for appropriate remedial measures within either country. The answer lies in better resource management. The operation of the Indus Treaty has been relatively smooth precisely because it has somehow been insulated, partially if not wholly, from the difficult political and military relationship between the two countries, i.e., from the 'security' angle. One must hope that this will continue.

India-Nepal/Bhutan

There are really no serious water security issues between India and Nepal or between India and Bhutan. What has been under discussion between India and these countries is the building of projects for the purpose of hydroelectric power, irrigation and flood moderation. The point to be noted in the present context is that there are no questions of security or insecurity here. There are some pending issues regarding water sharing between India and Nepal in relation to certain rivers, particularly in the context of the Mahakali Treaty of February 1996, but these are minor matters that do not warrant elaboration in this paper (Iyer, 1999b).

India-Bangladesh

It is only in the India-Bangladesh context that the expression 'water security' acquires some significance. Bangladesh is a water-abundant country, but it experiences seasonal shortages and seasonal excesses including floods. As the lowest riparian in the Ganga-Brahmaputra-Meghna (GBM) system, Bangladesh could well feel a certain sense of insecurity in relation to water, particularly considering the size of the upper riparian, India, in terms of land and population. Bangladesh is very conscious of the fact that there are 54 rivers (including rivulets and streams) crossing the Indo-Bangladesh border, and that 94 percent of the country's waters originate beyond its borders.

Clearly, Bangladesh needs some reassurances regarding water. Fortunately, the most contentious (and at one time seemingly intractable) issue between India and Bangladesh, the Ganga Waters Issue, which was a major component in the complex and difficult relationship between them, was resolved by the Ganges Water Sharing Treaty of December 1996. This treaty has a term of 30 years and is renewable beyond that period by agreement. After a shaky start because of low flows in the river in the 1997 lean period (Iyer, 1999a), the treaty has worked to the satisfaction of the two governments in the three succeeding lean periods (1998, 1999 and 2000). Water-sharing talks are proceeding regarding the Teesta River, on which both countries have built barrages. These talks need to be brought to a quick conclusion and followed by talks that relate to the other rivers, including the Muhuri, Manu, Gumti, Khowai, Brahmaputra, Dharla and Dudh Kumar.

Studies on Cooperation

The question of water security in South Asia has turned out to be a fairly limited one, requiring a few more agreements between India and Bangladesh. However, this may be dismissed as a reductionist view of security. Some, in India as well as in Bangladesh and Nepal, might argue that if we take a wider view of possibilities of cooperation among the countries of the GBM system, large vistas will open out before us. That may be true, but there is no particular reason why possibilities of cooperation, harmony and integrated development should be discussed under the rubric of 'security'. Nevertheless, this needs to be gone into a little further, as much work has been done on this theme already, and this writer has been associated with some of it.

More than ten years ago, at a time when there was an impasse regarding water resources at the intergovernmental level, either between India and Nepal or between India and Bangladesh, an effort was made to promote better understanding, cooperation and harmony through non-official initiatives. It took the form of a collaborative three-country study project on the possibilities of India-Bangladesh-Nepal cooperation on the utilization of the water resources of the GBM rivers. This was undertaken by the Centre for Policy Research of New Delhi, the Bangladesh Unnayan Parishad of Dhaka, and the Institute for Integrated Development Studies of Kathmandu. The first phase of this project resulted in four books: one by each of the institutions, leading to a 'consensus' book entitled *Converting Water into Wealth* (Verghese, et al., 1994) by all three institutions. The second phase of the project had a twofold objective: the dissemination and propagation of the ideas emerging from the first phase, and further work on some of the identified possibilities; this resulted in a book entitled *Cooperation on the Eastern Himalayan Rivers* (Adhikary, et al., 2000).

The basic thesis underlying the entire project was that the GBM region is a resource-rich area, but it is home to a large concentration of the world's poorest people; that the water resources of the immense rivers that flow through the region hold the promise of an escape from this paradox; that water is in fact the magic key to future prosperity in this region; and that the countries here must learn to cooperate to bring about that prosperity or else remain locked in grinding poverty.

Leaving aside the reservations that one has on this seemingly compelling rationale and wholly endorsing the plea for cooperation, one still needs to ask: in concrete terms, what does cooperation mean here? It means essentially cooperation in harnessing the water resources of the GBM system by means of a number of major projects such as the Karnali, Pancheswar and Sapt Kosi (three major tributaries of the Ganga) in Nepal; the Manas and Sankosh in Bhutan; the Dihang and Subansiri in India (on the Brahmaputra); the Tipaimukh in India (on the Barak); the Ganges Barrage in Bangladesh; and so on. Most of these would be multi-purpose projects with irrigation, hydropower and flood moderation components; some may also contain a navigation element.

There have also been a number of other studies on the subject of water. Apart from national studies and plans (for example, the Report of the National Commission on Integrated Water Resources Planning, submitted to the Government of India in September 1999; the National Water Plan of Bangladesh; etc.), there were a series of 'Water Vision

2025' exercises undertaken by different countries in South Asia under the auspices of the Global Water Partnership. These were held in preparation for the World Water Forum held at The Hague in March 2000. The 'Vision' exercises were partly national (as in India Water Vision and Pakistan Water Vision) and partly thematic (as in Water for Food and Water for Nature); these were eventually brought together into a 'South Asia Water Vision 2025' which was presented at the Hague Forum.

Scarcity and the Advocacy of Water Resource Development

Without going into the details of the studies, we may take note of a broad commonality of approach that characterizes many of these exercises, both national and regional. The general trend, with some exceptions, whether at the global, national, or regional level, is to argue that given the fast-growing demand for water arising from the growth of populations, the pace of urbanization, and the pursuit of higher standards of living, and given the finite nature of the supply of fresh water in nature, the world as a whole is likely to face a situation of a serious shortage of water within a decade or so, and that the stress from this shortage will be particularly severe in certain areas and regions. To that projection of a serious scarcity of fresh water the answer is thought to lie in a strong supply-side effort at 'water resource development' (WRD) in the form of large projects for the storage and transfer of river waters. Calculations are made of the large investments needed for WRD projects. Skepticism about the capabilities of governments to mobilize those resources or to manage the systems leads to the postulation of a large role for the private sector in WRD investments. Policy changes to facilitate such investments are advocated. That line of argument has been heard repeatedly during the last three years or so in numerous seminars, conferences, round tables and workshops, held under the auspices of governments, international organizations, universities, research institutions, think tanks, and others.

In this line of thinking the focus is on WRD, i.e., big projects. During the course of the deliberations at the numerous meetings that preceded the Hague Forum, a point repeatedly made by some (particularly engineers) was that water resource *development* was crucial and that the commonly-used phrase 'integrated water resource management' (IWRM) needed to be expanded to 'integrated water resources development and management' (IWRDM). In this advocacy of 'development', genuflections were duly made in the direction of environmental concerns, equity, social justice, and so on, but the central idea was WRD.

There are some unexamined ideas – partly unstated, partly explicit – underlying this argument for WRD. Firstly, demand projections are generally based on current patterns of water use with some adjustments for improvements in efficiency and resource conservation, or on prevailing notions of 'development'. If we assume, 'realistically', that no more than moderate improvements in efficiency and economy in water use are likely to happen, and if we assume further and (even more 'realistically') that there will be no change in our ideas of 'development' (i.e., the pursuit of higher and higher 'standards of living', with all countries aspiring to reach America's level), then there will be a horrendous water scarcity. We must then also 'realistically' acknowledge that sustainable development is an impossibility, and that planet earth is doomed. Instead of such apocalyptic forecasts, it will surely be more realistic to recognize that given the precious nature of this life-sustaining element and its finite supply, a tremendous effort needs to be made at efficiency, economy and conservation, so that the benefit derived from a unit of water is maximized.

The approach commonly used in the case of other consumer or industrial goods, that of projecting demand and providing the supply through production, is inappropriate in the case of water. Here we need to start from the recognition of a finite availability and learn to live with it. By way of an aside, one might add that in the course of the India-Bangladesh talks on the sharing of river waters, it has often been argued that even the waters of the mighty rivers Ganga and Brahmaputra are inadequate for the needs of the two countries. This would sound very strange to people in other parts of the world who have learned to live and live well with a far less generous natural endowment, Israel being an excellent example (Iyer, 1999b).

With the kind of reversal of approach indicated above (which implies changes in ways of living) water demand projections will undergo drastic reductions. Secondly, we must shake ourselves free of the usual engineering conventions of defining available water resources in terms of riverflows and usable water resources in terms of what is stored behind a dam. What is available in nature is rainfall, not just runoff, and while storing river waters behind a dam doubtless converts 'available' water into 'usable' water, so does *in situ* rainwater harvesting (i.e., 'catching the raindrop as it falls') and local watershed development. These are also part of the supply-side answers to demand.

If we combine those two propositions, namely, if we learn to reduce our draft on this finite and precious resource, and if we learn to augment supplies locally through watershed development and water-harvesting in every area where this is feasible, then we may find that the available water will go a long way indeed, and that a crisis can be averted with relatively little (if any) recourse to big WRD projects and huge investments.

Water Markets

If engineers and administrators tend to argue for supply-side projects, liberal economists and officials of the multilateral financial institutions tend to argue for water markets. To them, water is a commodity like any other, governed by the laws of supply and demand. If the state steps back and leaves it to the private sector and to the operation of market forces, then supply will meet demand, prices will correct, economy and conservation will be ensured, and the market will automatically resolve conflicts. The slogan is: 'water is an economic and social good'. Yes, water is an economic good when it is used for industry or agriculture, and perhaps a social good when used for sanitation or or for fighting fires, but is 'social good' an adequate description of water as a basic human and animal need and, indeed, as the sustainer of the environment of which it is a part? Can water in that basic aspect be reduced to a commodity like cement or steel or fertilizer or soap? Is it not more akin to the air we breathe? One is not ruling out water markets; they may have a role to play, but there are important issues of equity, social justice and sustainability that cannot be left to market forces. The glib answer to that will be that these can be taken care of through regulation, but regulation is far from easy.

The doctrinaire call for privatization includes allowing the corporate private sector to build and operate dams across rivers for hydroelectric power and/or for irrigation. Assuming that the private sector is interested in investing in such capital-intensive, long-term, modest-return projects, how are the environmental and social impacts going to be handled by the private entrepreneur and manager?

Alternatives

If we forget for a moment the questionable calculus of supply and demand and look at security from the point of view of protecting the ecological system and planet earth, we get a different perspective altogether; we begin to realize that by building a series of large WRD projects in this subcontinent we shall not be ensuring security but endangering it. Such projects constitute major interventions in nature, often in fragile ecosystems, and are fraught with incalculable risks.

It may be possible to mitigate or compensate for some of the impact, but much of it is irremediable. Severe disturbance of pristine areas, submergence of land and forests, reduction of biodiversity, radical alteration of the river regime, adverse impacts on aquatic life and estuarine conditions, uprooting and dispersal of communities and cultures, increased hazards in seismically active areas and other issues are among the unavoidable consequences of such projects. They may indeed create wealth, but they will not eliminate poverty; the wealth and the poverty will be in different hands. The alleviation of poverty, the ensuring of equity and social justice, the removal of the disabilities of women, children and disadvantaged groups and the empowerment of the community are hard tasks to be tackled with patience, imagination and determination *within* each of the countries of this region. A preoccupation with large WRD projects and the naive belief that they hold the magic key to prosperity will produce complacency and distract us from what needs to be done.

During the summer of 2000, newspapers in both India and Pakistan were full of reports about severe drought conditions in many parts of South Asia. These were not indications of water insecurity, nor did they point to the need for big projects or long-distance water transfers. Most of the comments, analyses and prescriptions that appeared in the media recognized that the drought conditions were the result of bad water management in the past and that the answer lay in better resource management in the future. Failure to harvest rainwater, excessive extraction of groundwater and failure to ensure the recharge of the aquifers led to the water table falling sharply over the years, so that when a bad year came there was no groundwater reserve to draw upon. That is a broad description of what went wrong, though conditions may have differed from place to place. That explanation is proven by the fact that in the same areas – for example, parts of Rajasthan and Gujarat in India – lush green villages can be found side by side with dry, brown villages; the former had practiced water harvesting and groundwater recharge for some years, the latter had not. It may be added that out of the 4291 large dams in India, the majority are in Maharashtra and Gujarat; the several hundred dams in Gujarat failed to provide drought-proofing when needed.

Thus, the answer to future demands for water – based neither on visions of excessive affluence nor of Gandhian austerity, but of modest prosperity – does not necessarily lie in large, centralized, top-down, technology-driven projects. Local, decentralized, community-based, people-centered alternatives are available and are likely to pay better dividends. In India, achievements such as those of Anna Hazare (Ralegan Siddhi village in Maharashtra), P.R. Mishra (Sukhomajri in Haryana), Rajendra Singh and the NGO Tarun Bharat Sangh (several hundred villages in Rajasthan), and similar efforts in Madhya Pradesh under the leadership of Chief Minister Digvijay Singh, have become well known. There must be similar examples in the other countries of the region, though this writer is not very familiar with them.

What needs to be understood is that these are not small instances but rather significant developments in terms of increased water availability, rise in groundwater tables, and the extension of irrigation. If these examples could be replicated in the thousands, all across South Asia, they could be major components in water planning; they are not to be patronizingly dismissed as 'minor', 'secondary' or 'supplementary to big projects', as often occurs. What also needs to be understood is that these are instances not merely of water management, but of social mobilization and transformation. Being community initiatives, they are generally more harmonious and equitable than large, centralized, top-down systems. Conflicts may arise even in such local efforts, but they are likely to be resolved within the community. For instance, the people involved in using the waters of the Arwari (a small river in Rajasthan) have established the 'Arwari Parliament' for decision-making on matters of common interest and for resolving conflicts. The answer to future needs lies in this kind of local, civil-society initiative, not in grandiose visions of WRD or long-distance transfers. This is the road to true sustainability; without sustainability there is no security.

Inter-Country Cooperation

The above does not serve to argue that there is no need for inter-country cooperation. The plea is merely that such cooperation should not be identified with a clutch of big projects or confined to the sphere of governments. There are many other possibilities and compulsions of cooperation. The protection of water sources (rivers, lakes, mountains, forests, and aquifers) from pollution, degradation or denudation; the preservation and regeneration of deteriorating wetlands; improving and maintaining water quality; dealing with common problems such as drainage in the Indus basin in India and Pakistan, or the occurrence of arsenic in aquifers in India and Bangladesh; coping with floods and minimizing damage; sharing experiences in local water harvesting and watershed development as well as in the related social mobilization and transformation – these are among the issues on which inter-country cooperation will be very fruitful, and in some instances outright necessary. Such cooperation can be at the level of governments, NGOs, academia or think tanks, or via people-to-people interaction.

Inter-country treaties or agreements regarding river systems that run along or straddle boundaries will, of course, remain necessary.

Conflicts may arise in the absence of such understandings, or even in relation to the interpretation of clauses or the actual operation of such agreements; mechanisms will be needed to resolve these conflicts. In this context it is interesting to note that conflicts over river waters, whether inter-country or intra-country, seem to arise particularly in the context of large projects.

For example, the India-Bangladesh dispute over Ganga waters was precipitated by the Farakka Barrage Project in India. On the Teesta the problem is that both India and Bangladesh have built barrages. On the proposed Tipaimukh Project on the Barak in Tripura in India, Bangladesh is apprehensive about possible impacts in its areas. Projects on the Kosi and Gandak were the starting points of a prolonged history of misunderstandings between India and Nepal, which was further accentuated by the Tanakpur Barrage Project, then it was resolved through the Mahakali Treaty. But even after that treaty there remain pending issues relating to the Pancheswar Project. Within India, the Cauvery dispute arose because dams and reservoirs built by Karnataka had the effect of reducing the flows into the Mettur Reservoir (an earlier project of Tamil Nadu); between Karnataka and Andhra Pradesh, there is a dispute over the Alamatti Project of the former on the Krishna River; also on the Krishna, the Telugu Ganga Project of Andhra Pradesh is the subject of a dispute between that state and Karnataka and Maharashtra. Even within a state, a large project creates conflicts of interest between the people of the upper catchment and those living downstream; between those who bear the social costs of the project and those who enjoy the benefits such as irrigation or hydro-power; between the head-reach farmers and those at the tail end of the canal system; between the rich farmers and the poorer ones in the command area; between human beings and wildlife; and so on. Such conflicts have been incidental to many projects, but have been particularly marked in the case of some. In India, the Sardar Sarovar Project on the Narmada and the Tehri Hydro-Electric Project on two tributaries of the Ganga in the Himalayan region have been the centers of fierce controversy and have become international *causes cèlébres*. In Nepal, Arun III became controversial, leading to the withdrawal of

World Bank support for it. There is also the continuing controversy raging around the Kalabagh Dam Project in Pakistan.

To put it provocatively, large projects appear to create or accentuate conflicts rather than resolve them. This is essentially because they involve major interventions in nature and in society. Naturally, they therefore become the foci of conflicts.

Conclusion

In conclusion, it may not be out of place to make a reference to two questions posed by a distinguished professor at a conference on the eastern Himalayan rivers at Kathmandu a few years ago. Referring to the sustainability debate, he said that the concerns involved could be captured in the following two questions: 'How far can we go in our draft on natural resources before we cause harm to future generations?' and 'How far can we go in continuing to dump pollutants and contaminants in water, soil and the atmosphere before we cause irreversible damage?' At first glance these questions seem well formulated, but if we consider them more carefully, we realize that they ask not a *positive* question ('how can we live in harmony with nature?') but a *negative* one ('how much harm and damage can we afford to inflict?'). Behind this lies an adversarial view of the relationship between humanity and nature, and the philosophy (still prevalent, though no longer so stridently voiced) of conquering nature. This is the legacy of the Western legend of Prometheus who is said to have brought fire to earth in defiance of the gods. Under the influence of that legacy, we are driven by technological hubris to undertake the 'harnessing' of nature for the sake of 'development'. In contrast, we have the Indian legend of Bhagiratha, who brought water – the river Ganga – to earth in a prayerful spirit.

This brings us back to the theme of security. True security lies in a healthy, or at least a non-pathological, relationship with nature. We cannot learn to live in harmony with one another until we have learned to live in harmony with nature.

Chapter Ten

Environment & Security in South Asia: Lessons & Conclusions

ADIL NAJAM

AS ALREADY DETAILED in Chapter 1, the purpose of this volume is neither to posit a new conceptual framework for understanding the interactions between environment and security nor to put any of the many existing frameworks to the test of empirical evidence emerging from South Asia.

This is a much more humble enterprise. The chapters in this volume have sought to enunciate the *various* ways in which environmental issues in South Asia tend to manifest themselves within the broader context of human security. The preceding chapters do not attempt to definitively uncover all of the very complex interactions between environment and security; they merely seek a better policy sense of how, in the context of the South Asian subcontinent, the twin challenges of environmental and human security are intertwined. The ultimate goal for the authors of this volume is not as much a search for causalities in the relationships between environment and security as it is a desire for policy coherence in what they see as the rather muddled praxis in both domains.

Written by authors who all come from South Asia and who bring a very wide variety of disciplinary perspectives — the writing team included those trained in economics, geography, engineering, system dynamics, physics, wildlife ecology, political science, urban planning, and more — the chapters in this volume distinguish themselves from the mainstream literature both in terms of their specific developmental orientation and their explicit policy focus. Indeed, the distinction between the language of these chapters and that of the mainstream literature is quite striking.

Contrary to the thrust of the mainstream literature which struggles, often unconvincingly, to express the environmental problematique in the language of state-centric 'national' security (for example, Mathews, 1989; Deudney, 1990; Homer-Dixon, 1991; Myers, 1993; Conca, 2001; Gleditsch, 2001), the chapters in this volume make it clear that in the context of South Asia (and arguably of most developing countries) it is the language of people-centric 'human' security that dominates the discussion of what might be considered simple environmental concerns elsewhere. This, more than anything else, may be the most important conclusion of this volume.

Poverty: A Key Link Between Environment and Security

The above notwithstanding, there are some important areas of concurrence between the findings of the preceding chapters and what we are calling the mainstream literature on environmental security. While the tone and focus of our chapters are markedly different, the issues that these chapters highlight and the conclusions that they arrive at often correspond to, and sometimes validate, the key lessons of the broader literature. While it is not our purpose to highlight all instances of such correspondence, two key findings from other scholars are so broadly validated in this volume that they deserve special mention.

The first relates to the models proposed by Thomas Homer-Dixon and his colleagues (see Homer-Dixon, 1991 and 1999; Homer-Dixon, Boutwell and Rathjens, 1993; Homer-Dixon and Blitt, 1998). An abridged version of the argument of this body of literature is presented in the introduction to the book *Ecoviolence* (Homer-Dixon and Blitt, 1998: 15):

> ... severe environmental scarcities often contribute to major civil violence. Poor countries are more vulnerable to this violence, because large fractions of their populations depend for their day-to-day livelihoods on local renewable resources.... Moreover, poor countries are often unable to adapt effectively to environmental scarcity because their states are weak, markets inefficient and corrupt, and human capital inadequate.

Various chapters in this volume seem to suggest the validity of this general finding (although often with caveats). In particular, the cases presented in the chapters on Pakistan, Sri Lanka and Nepal correspond with the thrust of this argument. Indeed, the links that these chapters draw between resource scarcity and ultimate civil strife are strongly supportive of the model posited by Thomas Homer-Dixon and suggest that such a linkage is relatively more likely in poor and weak states.

Related to the above is one of the key findings that Geoffrey Dabelko, Steve Lonergan and Richard Matthew (2000: 56) come to in their major literature review of the subject: "research on environment and security often strengthens the conclusion that poverty is a key factor in causing tension, unrest and, eventually, conflict." This finding is borne out quite strikingly by all of the chapters in this volume. Poverty emerges as the defining variable both in terms of defining environmental degradation and outlining human insecurity. Importantly, our authors identify poverty as the causal motivator of environmental stress as well as the most important manifestation of human insecurity.

While poverty is nearly always described as one of the key elements that exacerbate the causal chain that can lead from environmental degradation to violence and insecurity, the level of importance that the preceding chapters place on this variable suggests that it may well play a more central role in this chain of causality than much of the literature seems to acknowledge. Linking the two arguments described above within the context of the preceding chapters, one might argue that it is chronic and structural impoverishment, more than anything else, which forges the connection between environmental degradation and violent conflict. It is not simply that environmental degradation is *more likely* to lead to violent conflict in poor countries, but it is that poverty – in terms of economic, social or political disenfranchisement and vulnerability – may be a *required* condition for this connection to be made. This is an area that requires more empirical research and may, in fact, have the potential to untie many of the convoluted knots that the debate on the subject tends to find itself in.

Such a conception leads itself to a focus on social disruptions at the level of society (rather than the state). As posited in the framework presented in Chapter 1, this lends itself to conceptualizations related to human insecurity. This theme runs through nearly all the chapters of this volume, and particularly through the case studies presented in the country chapters. It should be noted that others who have looked at the myriad of security threats faced by South Asians have come to similar conclusions. Most notably, in launching the *Human Development in South Asia* reports, Dr. Mahbub ul Haq (1997: 84) identified the shift:

> Security is increasingly interpreted as: security of people, not just of territory; security of individuals, not just of nations; security through development, not through arms; security of all people everywhere – in their homes, on their jobs, in their streets, in their communities, and in the environment.

Five Key Lessons

In using a variety of contextual and analytical frameworks, the various chapters in this volume throw up a rather wide array of findings that are specific to their contexts, are not always comparable, and are sometimes contradictory. It is neither necessary nor particularly useful to repeat these specific findings in this concluding chapter. However, it is striking that despite the obvious contextual and analytical differences, there is a core set of conclusions that are common to most chapters or become obvious when one looks at all the chapters together.

The following five broad lessons are of particular importance because they have the potential to enrich the discussion on environment and security. These, of course, are to be understood within the context of the overarching conclusion already laid out above: a need to focus on poverty as a primary, but not sole, motor of human insecurity.

Lesson 1: For South Asia in particular, and developing countries in general, environment and security are best conceptualized within the context of Sustainable Development.

All the papers in this volume grapple, differently, with environmental and security issues in a decidedly human security framework which is most appropriately understood within a sustainable development context. Indeed, they tend to view human security as a fundamental requirement for the achievement of sustainable development. This is not entirely a surprise since the World Commission on Environment and Development (WCED, 1987) had itself made that connection explicit. It is nonetheless a noteworthy conclusion of this volume because so much of the literature on environment and security tends not to highlight this connection.

Placing the environment and human security problematique within the sustainable development complex has at least two important implications. First, such a conceptualization allows for articulating issues related to environment and security at the level of, and in the language of, policy and practice. Second, it contributes towards a better understanding of what sustainable development means in practice. Such a conceptual focus broadens the scope of the enquiry from whether and how environmental degradation might lead to societal and state insecurities to also include how human insecurities influence, or are influenced by, accelerated environmental degradation. Imperfect as it might be, sustainable development policy becomes a potential means of addressing the twin challenges of environmental degradation and human insecurity.

The chapters on India, Pakistan and Bangladesh are most explicit on drawing out the links between the environmental dimensions of human security and the broader policy goal of sustainable development. In all three chapters, the interplay of environmental degradation and human insecurity is articulated in the language of sustainable development; and, more importantly, in the language of sustainable development 'policy.' Sustainability emerges as an equally pivotal concept in other papers, especially those on energy and land resources. A key point is made in the chapter from Nepal, which highlights the fact that the notion of security is intricately related to the issue of livelihood and, therefore, cannot be de-linked from concerns about the content and context of the development experiment. Taken together, the chapters in this volume suggest that debates regarding human and environmental security are in themselves attempts to better understand and operationalize the concept of sustainable development.

On a related note, the authors of this volume have tended to ignore, if not resist and even reject, debates about whether environmental degradation should – or should not – be an element of 'traditional' military- and state-related security concerns (see, for example, Mathews, 1989; Homer-Dixon, 1991; Myers 1993; Levy 1995; Deudney and Matthew, 1999). To put it rather glibly, given their practitioner roots, most of our authors seem to find these 'hot' intellectual debates uninteresting if not irrelevant to the problems they are grappling. In this regard, the papers in this volume are a departure from the norm of the larger literature. For the most part, our authors seem more comfortable defining environmental security as one more component of sustainable development than as a dimension of 'traditional' national security.

Lesson 2: The challenge of environment and security in South Asia is principally a challenge at the domestic, even local, level; however, it is a challenge common to the region.

One of the conclusions reached by Dabelko, Lonergan and Matthew (2000: 56) in their review of the literature on the subject is that "the most severe challenges for individual well-being in many parts of the world may not be external (to the country of residence), but internal; although internal problems are likely to be affected in some way by external forces." The chapters in this volume echo and amplify this lesson. It is clear that the authors of this volume see the challenge of environment and security to be principally a problem that needs to be tackled at home.

Indeed, there is a distinct tendency to bring the problem *'down'* to the ground level rather than raise it *'up'* to national, let alone regional, levels. This is not to suggest that the chapters consider the regional dimensions to be unimportant; however, it does serve to underscore the view that local challenges are more numerous as well as more profound than the regional ones.

The chapter on water resources makes an explicit case against viewing environmental issues as problems of 'international security' – despite the fact that water is the most internationally shared of resources. Other chapters make their case for the need to focus on domestic and local dimensions more implicitly, but no less forcefully. Indeed, the chapter from Nepal goes further than most by questioning the centrality of the state even to issues of domestic environmental security. A key concept in this regard is that of 'multiple securities' as discussed in the chapter on Sri Lanka (and also highlighted in the chapter on India). The various case studies discussed in the chapters – for example, the Kalabagh Dam in Pakistan, the Pancheswar Dam in Nepal or the Mahaweli project in Sri Lanka – suggest that the most important policy questions tend to correspond to those points where multiple securities collide, either in terms of the different stakeholders or in terms of the different issues at stake.

The above should not suggest that looking at issues from a regional perspective is unimportant. Far from it, various chapters repeatedly make the point that the countries of the region can learn much from each other because there are so many similarities in the challenges they face. In particular, the chapters on crosscutting issues highlight the potential for cross-regional learning and cooperation. The chapter on energy, for example, charts the rather different trajectories followed by the countries of the region and highlights some of the areas where they can, in fact, learn from each others' experience. The chapter on land resources similarly highlights the fact that the nature of the challenges faced by the different countries is similar and there is much room for mutual learning. Many of the country studies further validate this view. However, the one lesson that all the papers seem to underscore is that the best thing that each country can do for the region is to clean up its own act at home.

Lesson 3: The challenge of environment and security in South Asia is, at its core, not just a problem of resource endowments or geography but, quite distinctly, a problem of institutions and governance; if only because it is the latter rather than the former that we have the ability to change.

Possibly our most significant conclusion relates to the centrality of institutions and governance to our understanding of how environment and security are linked in the South Asian context. At one level, the conclusions are not particularly surprising; it is quite well known, after all, that environmental crises in many developing regions, including South Asia, are not just crises of resources scarcity or degradation, rather that they are fundamentally tied to fragile institutions for environmental governance. This is a very important insight since so much of the literature on environment and security tends to underplay, if not outright ignore, the governance dimension.

In this very important respect, our authors are contradicting one of the major conceptual tenets of the larger literature on environmental security. While the literature tends to concentrate on resource scarcity and environmental degradation as the key defining variables, the authors of this volume tend to highlight weaknesses in institutions and governance as the key causal determinant. Indeed, many of the examples mentioned in the preceding chapters – including the Dir forestry case in Pakistan, the Tehri Dam in Nepal, or the Eppawala mining project in Sri Lanka – suggest that institutional and governance weaknesses can lead to significant human insecurity even in the absence of severe environmental scarcity or degradation. In these cases, and in others, it is the lack of appropriate institutions and governance that help explain not only the resulting human insecurity but also the scarcity and degradation that others tend to focus on.

The implication of the recommendations of the papers – and this is a critical point – is that the solutions to issues of environment and security will not come from techno-fixes and mega projects that might somehow 'overrule' the forces of geography and nature; the solutions are more likely to come from institutional and governance reform. The papers in this volume, particularly all of the country studies, have much to offer by way of explaining the type of institutions and governance structures that can help work to alleviate human insecurity as well as environmental stress. Importantly, they suggest that democracy counts, transparency counts, culture counts, decentralization counts, and most importantly, participation counts.

In broadening the focus beyond resource scarcity and degradation, the papers in this volume raise important conceptual issues for the literature on environment and security. An earlier generation of scholars had been preoccupied with the effects of security issues (particularly war and preparation for war) on the environment, or more precisely on natural resources (Galtung, 1982; Westing, 1984 and 1988; Renner, 1989). The current phase of interest in the subject has moved in the opposite direction and tends to focus on how environmental degradation can lead to insecurity and violence (Deudney and Matthew, 1999; Homer-Dixon, 1999; Diehl and Gleditsch, 2001). Both arguments have environmental conservation, rather than human security, as their core interest. In focusing on sustainable development as the metric of analysis and placing human security more centrally in the discussion, the preceding chapters point towards the importance of looking at both link together – the ways in which insecurity, at any level, impact the environment *and* the ways in which environmental scarcity and degradation impact insecurity, at any level.

Such a formulation also allows them to move away from the more restrictive notion of 'acute violence' to the more encompassing concept of insecurity, particularly human insecurity. In this regard, the conclusions of this volume echo a key finding of the larger literature: even where environmental variables do not directly cause conflict, they can increase insecurity by accentuating the variables that can precipitate conflict (Gurr, 1993; Libiszewski, 1992; Dabelko, Lonergan and Matthew, 2000). Importantly, they add that other forms of insecurity can accentuate the conditions conducive to environmental degradation, thereby increasing eventual environmental insecurity. The implication, of course, is that the key to understanding the link between environment and security in any given context may not lie in variables directly related to either (such as scarcity or war). It may lie, instead, in issues that impact but are not directly related to either, such as failure of institutions and governance.

Lesson 4: The prospects of inter-state violence in South Asia over environmental issues is slim; however, given the region's history of distrust and dispute, environmental differences can add to existing tensions and apprehensions and perpetuate the general sense of insecurity that pervades interstate relations in the region.

Unlike others who have studied environment and security in South Asia and who tend to consider the region as a prime 'action theatre' for environmental conflicts (Myers, 1989, 1993; Hassan, 1991, 1992), our crew of authors is far more careful about painting doom scenarios. Indeed, they agree that the prospects of outright war in South Asia over these issues are not high.

There is also a tone of caution that suggests that, given existing regional security tensions and apprehensions, the broader sense of insecurity that defines the region's interstate relations can be exacerbated by environmental concerns. Arguably, there are far more immediate causes of inter-state tension in the region. Environmental issues, even though they are important, are likely to be overwhelmed by these other more 'traditional' security concerns. Moreover, despite fractious relations, even India and Pakistan have demonstrated a remarkable degree of cooperation (and even occasional goodwill) in the shared management of a precious resource such as water. Indeed, the Indus Water Treaty remains one of the few areas of sustained cooperation between the two countries (this cooperation has been severely tested in recent months but, fragile as it is and despite much sabre-rattling, it remains intact).

This should not negate the importance of environmental degradation as a factor in regional security. Given the deep and strong security faultlines in the region, it is not at all inconceivable that cross-boundary environmental problems could *add* to an already volatile situation and exacerbate existing apprehensions and distrust. In his study of environment and security in South Asia, Norman Myers (1993: 117) posits a fundamental question: "How can we realistically suppose that environmental problems will not exert a substantial and adverse influence over the prospects for the region's security throughout the foreseeable future?" While our authors are far more concerned with the more immediate and local implications of environmental degradation and human insecurity they are not oblivious to the potential political repercussions. Indeed, all of the country chapters point out that the ultimate effect of human insecurity and environmental degradation tends to be political instability. The chapter on India, in particular, makes the causal link between human, national and regional security quite explicit. All of the country chapters would, therefore, agree with Shaukat Hassan (1991: 65) that "in South Asia environmental deterioration has a very direct and immediate impact on the economy of the states, which in turn affects social relations in ways detrimental to political stability."

Importantly, the country studies in this volume do view environmental deterioration as an important *additional stress* that is likely to deepen existing lines of conflict (rather than create new ones). This view is best articulated in the paper from India, which points out that "the traditional approach to security, with its primary focus on military power, must be broadened to encompass the real sources of insecurity — unsustainable development as an aftermath of environmental exploitation, resulting in a deepening of the traditional forms of conflict." This suggests that although the main challenge is domestic, the regional context cannot be ignored; largely because regional environmental disputes, even when they may be relatively minor in and of themselves, can flare into larger conflicts because they are played out in a context of general distrust and dispute.

Lesson 5: There is the potential – albeit small – for a new generation of security relations in South Asia emerging around the nexus of environment and security that is based on the principles of mutual trust, harmony and cooperation rather than on legacies of distrust and dispute.

The chapter on water makes the very important point that even though security (in the international context) is generally seen as an adversarial concept, the environment demands a politics of consensus and cooperation. On the one hand, this brings into question the utility of putting the two terms together. On the other hand, however, it points toward the potential for a new approach to security – one that stresses the need for cooperative management of shared environments rather than the adversarial contest over scarce resources. A number of chapters, especially including the one on India, stress the importance of this potential.

However, a cautionary note needs to be sounded about the potential for moving to a new generation of security relations that start from the necessity of cooperation rather than from a history of confrontations. Given the 'traditional' security profile of the region, it is unlikely that such cooperation would emerge of its own in the natural evolution of things. Indeed, even where the need for such cooperation is self-evident, the hurdles to its establishment are profound. One must note with some concern that there is surprisingly little discussion in the preceding chapters of the role that the South Asian Association for Regional Cooperation (SAARC) might play in fostering regional environmental cooperation. This demonstrates both the firm conviction of our authors that the key challenges for each of these countries are domestic rather than regional and the recognition that establishing meaningful cooperation between nations on the environment will be an uphill challenge and requires more than declaratory intent.

Yet, our authors do point out some initial steps that could begin creating an atmosphere of cooperation. An important first step could be the institutionalization of some level of region-wide information sharing and joint planning for common concerns such as water, climate change, biodiversity, etc. The sharing of best practices, particularly in the areas of technological and institutional innovations for environmental enhancement, is another obvious step that a number of chapters outline. A third step, which is not directly suggested in any paper but which is reflected in the very rationale of this volume, relates to the need to strengthen and deepen expert dialogue and communication at the regional level. Finally, the increasing prominence of global environmental politics and the North-South dimension within it is yet another reason for the developing countries as a whole, and regions such as South Asia within it, to think in terms of coalitional rather than individual politics related to the international environment (Najam, 1995 and 2000a; Najam and Sagar, 1998; Agarwal et al., 1999). The SAARC is, indeed, well placed to take such steps and should be urged to continue its efforts in this direction.

To end on a positive note, one might suggest that the environment has the potential to become an 'entry point' for wider regional cooperation. On the one hand, the very nature of the environmental problematique points towards the urgency of adopting a cooperative mindset. On the other hand, the language of human security at least allows for the potential of focusing on regional security without necessarily regurgitating stylized debates about traditional hurdles to cooperation. This, of course, is the most optimistic of our conclusions. But when one is dealing with the future of a billion and a half people, including some of the poorest and most vulnerable people in the world, one simply cannot afford to give up on optimism.

Meaningful regional cooperation for improved environmental and human security in South Asia may well be too much to hope for. But hope we must.

Acronyms & Abbreviations

AchE	acetlcholine esterase
ADB	Asian Development Bank
AMDP	Accelerated Mahaweli Development Programme
BAP	Biodiversity Action Plan
BARC	Bangladesh Agricultural Research Council
CDM	Clean Development Mechanism
CEB	Ceylon Electricity Board
DDT	di-chloro-diphenyl-trichloro-ethane
DPR	detailed project report
EDP	Energy Dependence Path
EEP	Energy Efficiency Path
EIA	Environmental Impact Assessment
EIP	Energy Impact Path
EKC	Environmental Kuznets Curve
ENSO	El Nino Southern Oscillation
FDC	Forest Development Corporation
GBM	Ganges-Brahmaputra-Meghna (system/delta)
GDP	gross domestic product
GEF	Global Environmental Fund
GHG	greenhouse gas
GNP	gross national product
GWh	gigawatt hours
HYV	high-yielding varieties
IPP	independent power projects

IWRM	Integrated Water Resource Management
IWRDM	Integrated Water Resources Development & Management
JVP	Janatha Vimukthi Peramuna
KKH	Karakorum Highway
KYWS	Kalkot Youth Welfare Society
LEDCO	Langjung Electricity Development Company
LPG	liquefied petroleum gas
MAF	million acre-feet
MTOE	million tonnes of oil equivalent
NATO	North Atlantic Treaty Organization
NBA	Narmada Bachao Andolan
NEAP	National Environmental Action Plan
NGO	nongovernmental organization
NWFP	Northwest Frontier Province
PPP	purchasing power parity
SAARC	South Asian Association for Regional Cooperation
SPM	suspended particulate matter
TCR	tons of coal replacement
TPES	total primary energy supply
UML	Nepal Communist Party Marxist-Leninists
UNCED	United Nations Conference on Environment and Development
UNCHE	United Nations Conference on the Human Environment
UNDP	United Nations Development Programme
UNEP	United Nations Environmental Programme
UNESCAP	United Nations Economic and Social Commission for Asia and the Pacific
UNIDIR	United Nations Institute for Disarmament Research
US/USA	United States/United States of America
WAPDA	Water and Power Development Authority
WB	World Bank
WHO	World Health Organization
WRI	World Resources Institute
WTO	World Trade Organization

Bibliography

Abramovitz, J.N. (1996). *Impoverished Future: The Decline of Freshwater ecosystems*. Worldwatch Institute, Washington, D.C.

Adede, A. (1992). 'International environmental law from Stockholm to Rio'. *Environmental Policy and Law*. 22:2.

Adhikary, K.B., Q.K. Ahmad, S.K. Malla, B.B. Pradhan, R. Khalilur, R. Rangachari, S. Rasheed, B.G. Verghese, (Eds.) (2000). *Cooperation on the Eastern Himalayan Rivers: Opportunities and challenges*. Konark Publishers, New Delhi.

Agarwal, A., S. Narain, A. Sharma. (1999). *The State of India's Environment: The Citizens' Fifth Report*. Centre for Science and Environment, New Delhi.

Agarwal, A., S. Narain, S. Sen, (Eds.) (1999). *The Fifth Citizen's Report on India's Environment*. Centre for Science and Environment, New Delhi.

Agarwal, B. (1995). 'Feminist Environmentalism or Eco-Feminism?' *Survey of the Environment,* The Hindu Publications, Chennai, India.

Ahmad, Q. (Ed.) (2000a). *Bangladesh Water Vision 2025: Towards a Sustainable Water World.* Bangladesh Water Partnership, Dhaka.

Ahmad, Q. (Ed.) (2000b). *Perspectives on Floods of 1998.* University Press Limited, Dhaka.

Ahmad, Q., A. Ahmed (2000). 'Social Sustainability, Indicators and Climate Change'. In *Climate Change and Its Linkages With Development, Equity, and Sustainability* by M. Munasinghe, R. Swart (Eds.). Intergovernmental Panel on Climate Change, The Hague.

Ahmed, A. (1999). 'Contamination of Groundwater with Arsenic: A Great Environmental Disaster in Bangladesh'. *Asia-Pacific Journal of Environment and Development* 5:2.

Ahmed, A. (2000). Speech presented to the International Conference on Bangladesh Environment, Dhaka, 14-15 January.

Alker Jr., H.R. Haas, P.M. Haas (1993). 'The Rise of Global Ecopolitics.' In *Global Accord,* N. Choucri (Ed.). MIT Press, Cambridge, Massachusetts.

Allauddin, M., C. Tisdell (1988). 'Impact of Agricultural Technology on the Instability of Food Grain Production and Yield: Data Analysis for Bangladesh'. *Journal of Development Economics* 26:7.

Anderson, D. (1994). *Energy, Environment and Economy: Complementaries and Conflicts in the Search for Sustainable Growth.* The World Bank, Washington, D.C.

Arjunas Consulting (1997). *Atlas of Sri Lanka.* Arjuna Consulting Company, Dehiwela.

Arrow, K., B. Bolin, R. Coatanza, P. Dasgupta, C. Folke, C. Holling, B. Jansson, S. Levin, K. Maler, C. Perrings, D. Pimentel (1995). 'Economic Growth, Carrying Capacity and the Environment'. *Science* 268.

Aryal, M. (1995). 'Dams: The Vocabulary of Protest.' *Himal* 8:4.

Azhar, R.A. (1993). *Commons, Regulation and Rent-seeking Behavior: Pakistan's Guzara Forests.* University of Chicago, Chicago.

Bach, N., K. Saeed (1992). 'Food Self-Sufficiency in Vietnam: A Search for a Viable Solution'. *System Dynamics Review* 8:2.

Bakhtiar, I. (1992). 'Killer on the Loose.' *The Herald*, April 1992.

Bandyopadhya, J. (1992). 'The Himalaya: Prospects for Constraints on Sustainable Development'. In *State of the World's Mountains*, P. Store, (Ed.). Zed Books, London.

Banuri, T. (1992). *Noah's Ark or Jesus' Cross?* Working Paper. Sustainable Development Policy Institute (SDPI), Islamabad.

Banuri, T., G. Hyden, J. Calestous, R. Marcia (1994). *Sustainable Human Development From Concept to Operation: A Guide for the Practitioner.* UNDP Discussion Paper. UNDP, New York.

BAP. (1998). *Biodiversity Action Plan for Pakistan.* Government of Pakistan, Islamabad, Pakistan.

BCAS. (1998). *Bangladesh 2020: A Long-Run Perspective Study.* University Press Limited, Dhaka.

Belgrave, R., C. Ebinger, H. Okino (Eds.) (1987). *Energy Security to 2000.* Westview Press, Colorado.

Bharara, R. (1999). 'Pasture Lands.' *Survey of the Environment.* The Hindu Publications, Chennai, India.

Bhatt, A. (1997). 'Choked Lungs'. *Survey of the Environment.* The Hindu Publications, Chennai, India.

Bhattacharya, M. (1997). 'Not a City of Joy.' *Survey of the Environment.* The Hindu Publications, Chennai, India.

Black, J. (1992). *What is Development?* Institute of Contemporary Studies Press, San Francisco.

Blakie, P. (1994). *At Risk: Natural Hazards, People's Vulnerability, and Disasters.* Routledge, New York.

Bowonder, B. (1981). 'The Myth and Reality of High Yield Varieties in Indian Agriculture,' *Development and Change* 12:2.

Bowonder, B. (1986). 'Deforestation in Developing Countries.' *Journal of Environmental Systems* 15.

Brandon, C. A. Hammond. (1995). *Valuing Environmental Costs in Pakistan.* Background paper for the 'Pakistan 2010' Report. Government of Pakistan, Islamabad, Pakistan.

Brock, L. (1991). 'Peace Through Parks: The Environment on the Peace Research Agenda.' *Journal of Peace Research* 28.

Buzan, B. (1991). *People, States and Fear* (2nd Ed.). Lynne Reinner, Boulder.

CEB. (1998). *Long-Term Generation Expansion Planning Studies.* Generation Planning Branch, Ceylon Electricity Board, Colombo.

CGG. (1995). *Our Common Neighborhood: Report of the Commission on Global Governance.* Oxford University Press, Oxford.

Chadha, D. (1999). *Environmental Advocacy in Government and Business*. Span, New Delhi.

Chandler, W. (1987). 'Designing Sustainable Economies'. In *State of the World 1987* by L. Brown, et al, W.W. Norton & Co., New York.

Clark, W. (1991). 'Energy and the Environment: Strategic Perspectives on Policy Design'. In *Energy and the Environment in the 21st Century*. MIT Press, Cambridge, Massachusetts.

Colinvaux, P.A. (1980). *The Fates of Nations: A Biological Theory of History*. Simon and Schuster, New York.

Conca, K. (2001). 'Environmental Cooperation and International Peace'. In *Environmental Conflict* by P.F. Diehl, N.P. Gleditsch (Eds.). Westview Press, Boulder.

CSE. (1999). *The State of India's Environment: the Citizens' Fifth Report*. Centre for Science and Environment, New Delhi.

Dabelko G., S. Lonergan, R. Matthew (2000). *State-of-the-Art Review on Environment, Security and Development Co-operation*. OECD Development Assistance Committee, Paris.

Dalby, S. (1992). 'Security, Modernity, Ecology: The Dilemmas of Post-Cold War Security Discourse'. *Alternatives* 17.

Dasgupta, P., K. Maler (1990). *The Environment and Emerging Development Issues*. The World Bank Annual Conference on Development Economics. The World Bank, Washington D.C.

Deudney, D.H. (1990). 'The Case Against Linking Environmental Degradation and National Security'. *Millennium: Journal of International Studies* 19:3.

Deudney, D.H. (1991). 'Environment and Security: Muddled Thinking.' *Bulletin of Atomic Scientists* 47:3.

Deudney, D.H., R.A. Matthew (Eds.) (1999). *Contested Grounds: Security and Conflict in the New Environmental Politics*. State University of New York Press, Albany.

Diehl, P.F., N.P. Gleditsch (Eds.) (2001). *Environmental Conflict*. Westview Press, Boulder.

Dixit, A. (2000). 'Water as an Agent of Social and Economic Change in Nepal'. In *Water for Food and Rural Development Approaches and Initiatives in South Asia*, P. Mollinga (Ed.). Sage, New Delhi.

Douglas, M. (1985). *Risk Acceptability According to the Social Sciences*. Routledge and Keegan Paul, London.

Dwivedi, O. (1997). *India's Environmental Policies, Programmes and Stewardship*. Macmillan Press Ltd., London.

Ebinger, C. (1990). 'Pakistan'. In *Energy & Security in the Industrializing World*, R. Thomas, B. Ramberg (Ed.). The University Press of Kentucky, Kentucky.

EDC. (1992). *Population Dynamics, Environmental Changes and Developing Processes in Developing Countries: Case Study from the Northern Areas of Pakistan*. Enterprise Development Consulting and UN Research Institute for Social Development, New York.

Ehrlich, P.R., A.H. Ehrlich (1970). *Population, Resources, Environment: Issues in Human Ecology*. Freeman, San Francisco.

EIA. (1998). *International Energy Outlook 1998 With Projections to 2010*. Energy Information Administration, Springfield, Virginia.

Elliott, L. (1998). *The Global Politics of the Environment*. New York University Press, New York.

Elliott, L. (2001). 'Regional Environmental Security: Pursuing a Non-Traditional Approach.' In *Non-Traditional Security Issues in Southeast Asia*, A.T. Tan, J.D.K. Boutin (Eds.). Select, Singapore.

Fernando, A., N. Denis (1980). *Major Ancient Irrigation Works of Sri Lanka, Volume XXII*. J.R. Asia, Colombo, Sri Lanka.

Fisher, D. (1993). *Non-military Aspects of Security: A Systems Approach*. Dartmouth Publishing Company, Brookfield, Vermont.

Fisher, J. (1998). *Nongovernments*. Kumarian Press, Hartford.

Forest Survey of India (1997). *The State of the Forest Report*. Forest Survey of India, Dehradun.

Gaan, N. (2000). *Environmental and National Security*. South Asian Publishers, New Delhi.

Gadgil and Vartak. (1976). 'Sacred groves of the Western Ghats in India', *Economic Botany* 30.

Galtung, J. (1982). *Environment, Development and Military Activity*. Columbia University Press, New York.

Ghimire, L. (2000). Comments following the Nepali Congress Central Water Resources Committee. *Kantipur*, 20 January.

Gizewski, P., T. Homer-Dixon (1998). 'The Case of Pakistan'. In *Ecoviolence: Links Among Environment, Population, and Security*, T.F. Homer-Dixon, J. Blitt (Eds.). Rowman and Littlefield, Lanham.

Gleditsch, N.P. (2001). 'Armed Conflict and the Environment'. In *Environmental Conflict*, P.F. Diehl, N.P. Gleditsch (Eds.). Westview Press, Boulder.

Gleditsch, N.P. (Ed.) (1998). *Conflict and the Environment*. Kluwer Press, Dordrecht.

Gleick, P. (1991). 'Environment and Security: The Clear Connections'. *Bulletin of the Atomic Scientists* April. 17-21.

Gleick, P.H. (1992). *Water and Conflict*. Project on Environmental Change and Acute Conflict. University of Toronto, Toronto.

Goldemberg, J., W. Reid (1999). *Promoting Development While Limiting Greenhouse Gas Emissions*. UNDP, New York.

GoP. (1998). *Compendium of Environmental Statistics of Pakistan, 1994-95*. Statistics Division, Government of Pakistan, Islamabad.

GoP. (1980). *Soil Survey of Pakistan, 1977-78*. Government of Pakistan, Islamabad, Pakistan.

Gunaratna, R. (1995). *Sri Lanka: A Lost Revolution*. Institute of Fundamental Studies, Sri Lanka.

Gurr, T.R. (1993). *Minorities at Risk: A Global View of Ethnopolitical Conflicts*. United States Institute of Peace, Washington, DC.

Gyawali, D. (2001). *Water in Nepal*. Himal Books, Kathmandu.

Gyawali, D., A. Dixit (1994). 'The Himalaya-Ganga: Contending with Inter-Linkages in a Complex System'. *Water Nepal* 4:1.

Gyawali, D., A. Dixit (2000). 'Mahakali Impasse.' in *Domestic Conflict and Crisis of Governability in Nepal*, D. Kumar (Ed.). Tribhuban University, Kathmandu.

Hafiz, A., A. Khan (1990). *Security of Small States*. Bangladesh Institute of International Strategic Studies, Dhaka.

Haider, R., A. Rahman, S.U. Haq (1992). *Cyclone '91 Revisited: A Perceptional Study*. Bangladesh Centre for Advanced Studies, Dhaka.

Halle, M. (2000). Foreword to *State-of-the-Art Review on Environment, Security and Development Co-operation* by G. Dabelko, S. Lonergan, R. Matthew. OECD, Paris.

Hanson, A. (1999). *In the Balance: Environment and Security in the NWFP*. CIDA-Cowater, Ottawa.

Hassan, S. (1988). 'Problems of Internal Stability in South Asia'. *PSIS Paper*. Graduate Institute of International Studies, Geneva.

Hassan, S. (1991). *Environmental Issues and Security in South Asia.* Adelphi Papers. International Institute for Security Studies, London.

Hassan, S. (1992). 'Environmental Sources of Conflict in the South Asian Subcontinent'. *Disarmament* 15:1.

Hinrichsen, D. (1998). 'Facing the Coming Water Crisis'. *Environmental Change and Project Report* Summer 1999.

Hollings, C. (1973). 'Resilience and Stability of Ecological Systems'. *Annual Review of Ecology and Systematics* 4.

Homer-Dixon, T. (1991). 'On the Threshold: Environmental Changes as Causes of Acute Conflict'. *International Security* 16:2.

Homer-Dixon, T. (1994). 'Environmental Scarcities and Violent Conflict: Evidence from Cases'. *International Security* 19:3.

Homer-Dixon, T.F. (1999). *Environment, Scarcity, and Violence.* Princeton University Press, Princeton, New Jersey.

Homer-Dixon, T.F., J. Blitt (Eds.) (1998). *Ecoviolence: Links Among Environment, Population, and Security.* Rowman & Littlefield, Lanham.

Homer-Dixon, T.F., J. Boutwell, G. Rathjens (1993). 'Environmental Change and Violent Conflict'. *Scientific American* 268:2. 38-45.

Huq, S., Z. Karim, M. Asaduzzaman, F. Mahtab (Eds.) (1998). *Vulnerability and Adaptation to Climate Change in Bangladesh.* Kluwer Academic Publishers, Dordrecht.

Hussain, A. (1990). 'Resource Depletion and Sustainable Development in South Asia'. Presented at the IUCN Convention, Perth.

ICHME. (1991). *Fact Sheet: Military Impact on the Environment and Development.* International Clearing House on the Military and the Environment, New York.

IEA. (1999). *Energy Statistics & Balances of Non-OECD Countries 1996-97.* International Energy Agency, Paris.

Iftekharuzzaman (1994). 'Security of Bangladesh: The Societal Aspects'. In *South Asia's Security Primacy of Internal Dimension*, Iftekharuzzaman (Ed.). Academic Publishers, Dhaka.

IMF. (1999). *International Financial Statistics Yearbook 1999.* International Monetary Fund, Washington, DC.

Inayatullah, C. (1996). *Improvements of Governance in Forestry.* Sustainable Development Policy Institute (SDPI), Islamabad.

Islam, M. (1999). 'National Energy Policy of Bangladesh and Discussion on Issues Related to Energy Security'. Presented at the National Defence College, Dhaka, 8 April.

Islam, N. (1994). 'Environment and Security in South Asia: A Regional Overview'. In *South Asia's Security Primacy of Internal Dimension* by Iftikharruzzaman (Ed.). Academic Publishers, Dhaka.

IUCN. (2000). *IUCN Red Book*. IUCN-The World Conservation Union, Gland, Switzerland.

Ives, J., B. Messerli (1989). *The Himalayan Dilemma*. Routledge, London.

Iyer, R.R. (1998). 'Scarce Natural Resources and Language of Security.' *Economic and Political Weekly*, 16 May.

Iyer, R.R. (1999a). 'Conflict Resolution: Three River Treaties'. *Economic and Political Weekly*, 12 June.

Iyer, R.R. (1999b). 'The Fallacy of 'Augmentation''. *Economic and Political Weekly*, 14 August.

Jairajpuri, M. (1991). 'Habitat, the Change that Destroys.' *Survey of the Environment*. The Hindu Publications, Chennai, India.

Jalal, K. (1993). *Sustainable Development, Poverty and Environment Nexus*. Occasional Paper, Asian Development Bank, Manila.

Jodha, N. (1986). 'Common Property Resources and Rural Poor'. *Economic and Political Weekly*, 21:27.

Jodha, N. (1998). *Poverty-Environmental Resource Degradation Links: Questioning the Basic Premises*. International Center for Integrated Mountain Development, Kathmandu.

Jones, R.W. (1995). 'Message in a Bottle? Theory and Praxis in Critical Security Studies'. *Security Policy* 16:3.

Josse, M. (1994). 'The Case for New Thinking'. *Water Nepal* 4:1.

Kaplan, R. (1994). 'The Coming Anarchy'. *Atlantic Monthly*, February.

Karim, Z., A. Ibrahim, A. Iqbal, M. Ahmed (1990). 'Drought in Bangladesh: Agriculture and Irrigation Schedules for Major Crops'. Bangladesh Agricultural Research Council, Dhaka.

Karim, Z., S. Hussain, M. Ahmed (1990a). *Salinity Problems and Crop Intensification in the Coastal Regions of Bangladesh*. Bangladesh Agriculture Research Council, Dhaka.

Karve, I. (1961). *Hindu Society*. Deccan College, Pune.

Khan, H.A. (1997). *Implementing NEQS: Pakistan's Experience in Industrial Pollution Control.* Working Paper, Sustainable Development Policy Institute (SDPI), Islamabad, Pakistan

Khan, S. (1998). *Environment: A People's Perspective.* Paper prepared for the Third Sustainable Development Conference, Sustainable Development Policy Institute (SDPI), December.

Khan, S. (1999). *The Case Against Kalabagh Dam.* Sustainable Development Policy Institute (SDPI), Islamabad.

Khan, S., Naqvi, A. (2000). *The Environment-Poverty Nexus.* Sustainable Development Policy Institute (SDPI), Islamabad.

Knudsen, A. (1995). *Forest Management in Pakistan: Failed Policies or Local Mismanagement.* Michelson Institute of Development Studies and Human Rights, Bergen.

Kumarmangalam, P. (1998). Rajya Sabha (Indian Parliament), Response to Question No. 1600. Rajya Sabha, New Delhi.

Lama, H. (1997). 'Economic Resources and Environmental Concern in South Asia: Changing Interface,' in *Sustainable Development* (Ed.). Macmillan India, New Delhi.

Leitzmann, M., D. Vest. (1999). 'Environment and Security in an International Context'. *Environmental Change and Security Project Report* 41.

Levy, M. (1995). 'Is the Environment a National Security Issue?' *International Security* 20:2.

Libiszewski, S. (1992). *What is an Environmental Conflict?* Center for Security Studies and Conflict Research, Zurich.

Lodgaard, S. (1992). 'Environmental Security, World Order and Environmental Conflict. In *Conversion and the Environment* by N.P. Gleditsch (Ed.) Peace Research Institute, Oslo.

Mahbub ul Haq (1997). *Human Development in South Asia, 1997.* Oxford University Press, Karachi.

Maniruzzaman, T. (1982). *The Security of Small States in the Third World.* Australian National University, Canberra.

Mansingh, S. (1982). *India's Search for Power.* South Asian Publishers, New Delhi.

Mathews, J.T. (1989). 'Redefining Security'. *Foreign Affairs* 68:2.

Mathews, J.T. (1997). 'Power Shift'. *Foreign Affairs* 76:1.

McCully, P. (1998). *Silenced Rivers: The Ecology and Politics of Large Dams.* Orient Longman, Hyderabad.

McMichael, A.J. (1993). *Planetary Overload.* Cambridge University Press, Cambridge.

MEF. (1992). *Traditions, Concerns and Efforts in India.* Report to UNCED, Ministry of Environment and Forests, New Delhi.

Ministry of Finance (1999). *Survey of Bangladesh's Economy, 1999.* Government of the People's Republic of Bangladesh, Dhaka.

Ministry of Forestry and Environment (1995). *Sri Lanka Forestry Sector Master Plan.* Government of Sri Lanka, Colombo.

Ministry of Forestry and Environment (1999). *Database of Municipal Waste in Sri Lanka.* Government of Sri Lanka, Colombo.

Ministry of Information and Broadcasting (2000). *India 2000: A Reference Manual.* Government of India, New Delhi.

Ministry of the Environment and Forests (MEF) (1995). *Forestry Master Plan.* Government of the People's Republic of Bangladesh, Dhaka.

Ministry of Water Resources (1999). *Government of India Report.* Ministry of Water Resources, New Delhi.

Mirza, M. (1998). 'Diversion of the Ganges Water at Farakka and its Effects on Salinity. *Environmental Management* 22:5.

Moench, et al. (1999). *Rethinking the Mosaic: Investigations into Local Water Management.* Water Conservation Foundation, Kathmandu.

Mohnot, M. (1999). 'Mining.' *Survey of the Environment.* The Hindu Publications, Chennai, India.

Mollinga, P. (2000). *Water for Food and Rural Development.* Sage Publications, New Delhi.

Moomaw, W., G. Unruh (1997). *Are Environmental Kuznets Curves Misleading Us?* Tufts University, Medford.

Munasinghe, M. (1993a). *Environmental Economics and Sustainable Development.* World Bank, Washington, DC.

Munasinghe, M. (1993b). 'The Economist's Approach to Sustainable Development'. *Finance and Development* 30:4.

Munasinghe, M. (1995a). *Sustainable Energy Development (SED): Issues and Policy.* World Bank, Washington, DC.

Munasinghe, M. (1995b). 'Making Economic Growth More Sustainable'. *Ecological Economics* 15:2.

Munasinghe, M. (1999a). *Electric Power and Sustainable Development.* Presented at the IAEA General Conference, Vienna.

Munasinghe, M. (1999b). 'Is Environmental Degradation an Inevitable Consequence of Economic Growth: Tunneling Through the Environmental Kuznets Curve.' *Ecological Economics* 29.

Munasinghe, M. (2000). *Development, Equity and Sustainability.* Intergovernmental Panel on Climate Change, Geneva.

Murthy, V. (1978). *Environmental Problems of Water Resource Development in the Himalayan Region.* Proceedings of a National Seminar, New Delhi.

Myers, N. (1989). 'Environmental Security: The Case of South Asia.' *International Environmental Affairs* 1:2.

Myers, N. (1993). *Ultimate Security: The Environmental Basis of Political Stability.* W.W. Norton, New York.

Myers, N.. (1991). 'The Biodiversity Challenge: Expanded 'Hotspots' Analysis'. *Environmentalist* 10:4.

Nadkarni, V. (1997). 'Poverty, Dependence on Common Property, Land Resources and Economic Development'. In *Sustainable Regeneration of Degraded Lands*, J.L. Parikh Jyoti, S. Reddy (Eds.). Tata-McGraw Publishing, New Delhi.

Najam, A. (1993). 'Integrating Environmental Negotiation: A Strategy for the South'. In *Papers on International Environmental Negotiation*, L. Suskind, W. Moomaw, A. Najam (Eds.). The Program on Negotiation at Harvard Law School, Cambridge, Mass.

Najam, A. (1994). 'The South in International Environmental Negotiations'. *International Studies* 31:4.

Najam, A. (1995). 'An Environmental Negotiation Strategy for the South'. *International Environmental Affairs* 7:3.

Najam, A. (1996). 'Understanding the Third Sector: Revisiting the Prince, the Merchant and the Prince'. *Nonprofit Management and Leadership* 7:2.

Najam, A. (1999). 'Citizen Organizations as Policy Entrepreneurs'. *International Perspectives on Voluntary Action: Reshaping the Third World,* David Lewis (Ed.). Earthscan, London.

Najam, A. (2000a). 'The Four C's of Third Sector-Government Relations: Cooperation, Confrontation, Complementarity, and Co-optation'. *Nonprofit Management and Leadership* 10:4.

Najam, A. (2000b). 'The Case for a Law of the Atmosphere'.

Atmospheric Environment 34:23.

Najam, A. (2000c). 'Trade and Environment After Seattle.' *Journal of Environment and Development* 9:4.

Najam, A., A. Sagar (1998). 'Avoiding a COP-out: Moving Towards Systematic Decision-Making Under the Climate Convention'. *Climatic Change* 39:4.

Nakicenovic, N., A. Grubler, A. McDonald (Eds.) (1998). *Global Energy Perspectives.* Cambridge University Press, Cambridge.

NARESA. (1991). *Natural Resources of Sri Lanka: Conditions and Trends.* Government of Sri Lanka, Colombo.

Natarajan, S. (1990). 'Firewood Balance in the Nineties.' *Margin.* National Council of Applied Economic Research, New Delhi.

NEAP. (1998). *National Environmental Action Plan, 1998-2001.* Ministry of Forestry and Environment, Colombo.

Nelson, D. (1999). *Environment, No Time to Waste.* Span, New Delhi.

Okita, S., et al. (1989). *The Asian Development Bank in the 1990s: Report of a Panel.* Asian Development Bank, Manila.

Opschoor, J.B. (1989). 'North-South Trade, Resource Degradation and Economic Security'. *Bulletin on Peace Proposals* 20:2.

Panayotou, T. (1993). *Empirical Tests and Policy Analysis of Environmental Degradation at Different Stages of Economic Development.* International Labor Office, Geneva.

Panday, B. (1996). 'Local Benefits from Hydro Development'. *Studies in Nepali History and Society* 1:2.

Panday, K. (1998). 'Hydropower in Nepal'. *Water Nepal* 6:1.

Parikh, J., Reddy, S. (1997). *Sustainable Regeneration of Degraded Lands.* Tata-McGraw Publishing, New Delhi.

Pearson, P. (1994). 'Energy Externalities and Environmental Quality: Will development cure the ills it creates?' *Energy Studies Review*, 6.

Perceival, V., Homer-Dixon, T. (1995). 'The Case of the Northeast Region. *Project on Environment, Population and Security.* University of Toronto, Toronto.

Perrings, C., Opschoor, H. (1994). 'The Loss of Biodiversity: Some Implications'. *Environment and Resource Economics* 4.

Pezzey, J. (1989). *Economic Analysis of Sustainable Growth and Sustainable Development.* The World Bank, Washington, DC.

PFRI. (2000). *Provincial Forest Resource Inventory.* Government of the Northwest Frontier Province, Peshawar, Pakistan.

Pimm, S. (1984). 'The Complexity and Stability of Ecosystems'. *Nature* 307.

Porter, G. (1995). *Environmental Security as a National Security Issue.* Environmental and Energy Study Institute, Washington, DC.

Porter, G., J.W. Brown, P.S. Chasek (Eds.) (2000*). Global Environmental Politics (3rd Ed.).* Westview Press, Boulder.

Postel, S. (1999). *Pillar of Sand: Can The Irrigation Miracle Last?* W.W. Norton, New York.

Potkin, A. (1987). 'Towards Watering the Bangladesh Sundarbans'. *Proceedings of the Fifth Symposium on Coastal and Ocean Management*, Seattle.

Princen, T., M. Finger, J. Manno (1995). 'Nongovernmental Organizations in World Environmental Politics.' *International Environmental Affairs* 7:1.

Rahman, M. (1994). 'Environmental Aspects of Soils'. In *Environment and Development in Bangladesh, Vol. 1*, A. Rahman, R. Haider, S. Huq, E. Jansen (Eds.). University Press Limited, Dhaka.

Rashid, H., B. Paul (1987). 'Flood Problems in Bangladesh: Is There an Indigenous Solution?' *Environment Management* 11:2.

Renner, M. (1989). 'The Military's War on the Environment.' *State of the World 1989* by L. Brown et al. W.W. Norton, New York.

Renner, M., M. Pianta, C. Franchi (1991). 'International Conflict and Environmental Degradation'. In *New Directions in Conflict Theory,* R. Vayryned (Ed.). Sage Publishers, London.

Repetto, R. (1994). *The Second India: Revisited.* World Resources Institute, Washington, D.C.

Richardson, G. (1997). *Modeling for Management, Vol. II.* Dartmouth Publishing Company, Brookfield, Vermont.

Roy, A. (1999). *Cost of Living.* The Modern Library, New York.

Saeed, K. (1994). *Industry and Environment in Asia and the Pacific Region.* UNESCAP, Bangkok.

Schwartz, D.M., T. Deligiannis, T. Homer-Dixon (2001). 'The Environment and Violent Conflict'. In *Environmental Conflict,* P.F. Diehl, N.P. Gleditsch (Eds.). Westview Press, Boulder.

SDPI. (1995). *Nature, Power, People: A Citizen's Report*. Sustainable Development Policy Institute (SDPI), Islamabad.

Seminar (1999). 'Floods: A Symposium on Flood Control and Management'. *Seminar* 478.

Sen, A. (1984). *Resources, Values and Development.* Blackwell, Oxford.

Shelley, M. (1986). 'Nation-Building and Political Development in Bangladesh'. In *Nation Building in Bangladesh,* M. Hafiz, A. Khan (Eds.). Bangladesh Institute of Strategic Studies, Dhaka.

Shukla, A., A. Vandana (Eds.) (1999). *Advances in Environmental Bio-Pollution.* Ashish Publications, New Delhi.

Singh, J. (1999). *Pakistan's Fourth War for Kashmir.* Knowledge World, New Delhi.

Sinha, R. (1998). *Human Ecology*. Inashree Publishers, Jaipur.

SOE. (2002). *State of the Environment Report 2002*. Mahaweli Authority of Sri Lanka, Colombo.

Soroos, M. (1991). *Global Change, Environmental Security, and the Prisoner's Dilemma.* Prepared for the XVth World Congress of the International Political Science Association, Buenos Aires, July.

Soroos, M.S. (1994). 'Global Change, Environmental Security, and the Prisoner's Dilemma'. *Journal of Peace Research* 31:3.

Soroos, M.S. (1997). *The Endangered Atmosphere: Preserving a Global Commons.* University of South Carolina Press, Columbia.

Sterman, J. (2000). *Business Dynamics, Systems Thinking and Modeling for a Complex World.* Irwin McGraw-Hill, New York.

Stern, D. (1996). *Progress on the Environmental Kruznets Curve?* The Australian National University, Canberra.

Stone, C.D. (1993). *The Gnat is Older than Man: Global Environment and Human Agenda*. Princeton University Press, Princeton.

Strong, M. (1992). 'Energy, Environment and Development'. *Energy Policy* 20:6.

Suhrke, A. (1999). 'Human Security and the Interest of States'. *Security Dialogue* 30:3.

Susskind, L.E. (1994). *Environmental Diplomacy: Negotiating More Effective Global Agreements.* Oxford University Press, New York.

Swain, A. (1996). *The Ganges River Diversion, Bangladeshi Migration and Conflicts in India.* Upsala University, Sweden.

Thomas, C. (1992). *The Environment in International Relations.* The Royal Institute of International Affairs, London.

Thomas, R. (1990). 'India.' In *Energy and Security in the Industrializing World* by R. Thomas, B. Ramberg. The University Press of Kentucky, Kentucky.

Thomas, R., B. Ramberg (1990). *Energy and Security in the Industrializing World.* The University Press of Kentucky, Kentucky.

Thompson, K. (1992). 'Lowering Sights on the Road to Rio'. *IIED Perspectives* 8.

Thompson, M. (1997). *An Anti-Reductionist Framework for Thinking About Relationship Between Us and Rest of Nature.* Norwegian Research Centre in Organisation and Management, Bergen.

Ullman, R. (1983). 'Redefining Security'. *International Security* 8.

UNCHS. (1987). *Global Report on Human Settlements 1986.* Oxford University Press, London.

UNDP. (1994). *Human Development Report, 1994.* Oxford University Press, Oxford.

UNDP. (1999). *Human Development Report 1999.* Oxford University Press, New York.

UNDP. (2000). *Human Development Report 2000.* Oxford University Press, New York.

UNESCAP. (1986). Environmental and Socio-Economic Aspects of Deforestation in Asia and Pacific. United Nations Economic and Social Commission for Asia and the Pacific, Bangkok.

United Nations Security Council (1992). *Note by the President of the United Nations Security Council*, UN document S/23500. United Nations, New York.

Vandana A. (1999). *India's Foreign Policy and Subcontinental Politics.* Kanishka Publishers and Distributors, New Delhi.

Vandana, A. (1992). *The Politics of the Environment.* Ashish Publications, New Delhi.

Verghese, B. (1990). *Waters of Hope.* Oxford and IBH, New Delhi.

Verghese, B.G., R.R. Iyer, Q.K. Ahmad, S.K. Malla, B.B. Pradhan (Eds.) (1994). *Converting Water into Wealth.* Konark Publishers, New Delhi.

Wallingford, H. (1995). 'Sedimentation Studies in the Upper Mahawali Catchment.' Reprinted in *State of the Environment Report 2002.* Mahaweli Authority of Sri Lanka, Colombo.

Walt, S.M. (1991). 'The Renaissance of Security Studies'. *International Studies Quarterly* 35:2.

WCED. (1987). *Our Common Future: Report of the World Commission on Environment and Development.* Oxford University Press, London.

Westing, A.H. (1984). *Environmental Warfare.* Taylor and Francis, London.

Westing, A.H. (1988). *Global Norms, War and the Environment.* Oxford University Press, Oxford.

World Bank. (1992). *World Development Report, 1992: Development and the Environment.* The World Bank, New York.

World Bank. (1995). *Pakistan Poverty Assessment.* Country Operations Division, South Asia region, World Bank, Washington, D.C.

World Bank. (1997). *A Review of Key Environmental Issues in Bangladesh.* The World Bank, Dhaka.

World Bank. (1998). *World Bank Report 1998.* World Bank, New York.

World Bank. (2000). *World Development Report 2000/2001: Attacking Poverty.* Oxford University Press and World Bank, New York.

WRI. (1997). *World Resources 1995/96.* World Resources Institute and Oxford University Press, Oxford.

WRI. (2000). *World Resources 1998/99.* World Resources Institute and Oxford University Press, Oxford.

Index

Contributors

AHSAN U. AHMED heads the water and environment division of the Bangladesh Unnayan Parishad, a multi-disciplinary research center in Dhaka. He is an active policy researcher who has written on a host of issues related to environmental policy in Bangladesh, in particular on issues pertaining to global climate change.

VANDANA ASTHANA teaches political science at Christ Church College, Kanpur University, India. Her research focuses on environmental politics, environmental history and environmental security in India. She is author of the book *Politics of the Environment: A Profile.*

ZAHID H. CHOWDHURY teaches political science at the University of Dhaka, Bangladesh. His research focuses on corruption study, and socio-economic and political perspectives of environmental and natural resources management in Bangladesh.

AJAYA DIXIT taught hydraulics at the Institute of Engineering in Nepal for twelve years and has been an independent researcher since 1989. He is Editor of the journal *Water Nepal* and has also served two terms as a member of Nepal's Water and Energy Commission.

Kumudu Gunasekera is a doctoral candidate at the Department of Geography at Boston University. His research focuses on issues related to energy, infrastructure and the environment in developing countries.

Dipak Gyawali has been working with the water sector since 1979, initially as a government engineer and since 1987 as an independent analyst. His research focuses on the interface between technology and society, mainly around issues of water and energy. He is a Pragya (member) of the Royal Nepal Academy of Science and Technology and co-editor of the interdisciplinary journal *Water Nepal.*

Ramaswamy R. Iyer is a former Secretary for Water Resources, Government of India. He has been a Research Professor at the Center for Policy Research in New Delhi, a member of India's National Commission on Integrated Water Resources Planning, and a consultant to the World Commission on Dams. He is currently a member of the Vision 2020 Committee of India's Planning Commission and a Senior Advisor to Winrock International (India).

Shaheen Rafi Khan is a visiting research fellow at the Sustainable Policy Development Institute, a policy think tank based in Islamabad, Pakistan. He has a Ph.D in Economics from Columbia University. The present focus of his research is on the social dimensions of environmental management.

Sarath W. Kotagama was trained as an Ornithologist and Wildlife Ecologist and is now a professor of zoology and environmental studies at the University of Colombo, Sri Lanka. He has over 30 years of experience as an environmental NGO activist, a government policy-maker and an ecological scholar.

Adil Najam teaches international relations and environmental policy at Boston University and is a visiting fellow at the Sustainable Development Policy Institute in Islamabad, Pakistan. He is co-author (with Tariq Banuri) of *Civic Entrepreneurship: A Civil Society Perspective on Sustainable Development.* He is a winner of the MIT

Goodwin Medal for Excellence in Teaching and the Stein Rokan Award of the International Political Science Association.

A. ATIQ RAHMAN is the Executive Director of the Bangladesh Centre for Advanced Studies. He is a coordinator of the North-South Dialogue on Climate Change, convenor of the Climate Action Network South Asia, and coordinator of the Global Forum on Environment and Poverty. He has taught and researched at leading international universities, including Oxford University, Aberdeen University, Tufts University and the Massachusetts Institute of Technology.

KHALID SAEED, originally from Pakistan, heads the Social Science and Policy Studies department at the Worcester Polytechnic Institute in USA. He received Jay Wright Forrester Award for his work on sustainable development in 1995 and is widely recognized for his work on computer modeling and experimental analysis of developmental, organizational and governance-related problems.

ASHOK C. SHUKLA teaches political science at Christ Church College, Kanpur University, India. His research focuses on South Asia studies in general and on environmental security in South Asia in particular. He is editor of *Advances in Environmental Biopollution*.

The Regional Center for Strategic Studies

This book is the product of a South Asia wide research project of the Regional Centre for Strategic Studies (RCSS). RCSS is an independent, non-profit and non-governmental organization for collaborative research, networking and interaction on strategic and international issues pertaining to South Asia. Set up in 1992, the RCSS is based in Colombo, Sri Lanka.

The RCSS is a South Asian forum for studies, training and multi-track dialogue & deliberation on issues of regional interest. All activities of RCSS are designed with a South Asia focus and are usually participated by experts from all South Asian countries. The Centre is envisaged as a forum for advancing the cause of co-operation, security, conflict resolution, confidence building, peace and development in the countries of the South Asian region.

The RCSS serves its South Asian and international constituency by: a) networking programs that promote interaction, communication and exchange between institutions and individuals within and outside the region engaged in South Asian strategic studies; b) organizing regional workshops & seminars and sponsoring and coordinating collaborative research; and c) disseminating output of the research through publications which include books, monographs and a quarterly newsletter. The RCSS facilitates scholars and other professionals of South Asia to address, mutually and collectively, problems and issues of topical interest for all countries of the region.

Queries may be addressed to:

Regional Centre for Strategic Studies
2, Elibank Road
Colombo 5, SRI LANKA.
Tel: (94-1) 599734, 599735; Fax: 599993
e-mail: rcss@sri.lanka.net
RCSS website: http://www.rcss.org